AN INTRODUCTION TO
SAGE PROGRAMMING

AN INTRODUCTION TO SAGE PROGRAMMING

With Applications to SAGE Interacts for Numerical Methods

RAZVAN A. MEZEI
Lenoir-Rhyne University
Donald & Helen Schort School of Mathematics and Computing Sciences
Hickory, NC
USA

Published by John Wiley & Sons, Inc., Hoboken, New Jersey
Published simultaneously in Canada

For general information on our other products and services or for technical support, please contact our Customer Care Department within the United States at (800) 762-2974, outside the United States at (317) 572-3993 or fax (317) 572-4002.

Wiley also publishes its books in a variety of electronic formats. Some content that appears in print may not be available in electronic formats. For more information about Wiley products, visit our web site at www.wiley.com.

Library of Congress Cataloging-in-Publication Data applied for.

ISBN:9781119122784

Typeset in 10/12pt TimesLTStd by SPi Global, Chennai, India.

Printed in the United States of America

10 9 8 7 6 5 4 3 2 1

1   2016

I would like to dedicate this work to all my instructors who passionately directed my interest into the great fields of Mathematics, Computer Science, and Statistics.

To name just a few of them, in chronological order: Aurel Netea, Petru Dragos, Barnabas Bede, Alina Alb Lupas, Dan Noje, Mircea Balaj, Sorin Gal, Mircea Dragan, Alexandru Bica, Ioan Fechete, Ioan Dzitac, George Anastassiou, Maria Botelho, James Campbell, E. Olusegun George, Nikos Frantzikinakis, and not the least Seok Wong. There are many more, but the named ones modeled my thinking and gave me a direction to where I am today and influenced me the most.

Thank you from all my heart for your dedication, support, and friendship.

RAZVAN A. MEZEI

CONTENTS

PREFACE

This work is intended to be a gentle introduction to programming in Sage Math and Sage Interacts. It assumes no programming background from the reader, and it is specifically tailored for Mathematics, Mathematics Education, and Engineering students and instructors.

The book starts with a description on how one can use Sage Math as a calculator. It also explains how one can use it for computations and for plotting. Then, it covers a brief and gentle introduction to programming in Sage Math. You will learn how to create your own methods in Sage Math and how to create Sage Interacts. The book ends with a chapter that gives several examples on how one can use Sage Interacts for various Numerical Methods.

If you have no programming background yet, your programming skills need some improvements, you want to learn how to use Sage Math to program some numerical methods, or you want to create neat interactive representations of some mathematical concepts, then this book is for you.

The book, however, does not address in detail the mathematical topics covered in the given Sage Interacts examples. In particular, no mathematical proofs are given in here. If you want to study the Mathematical side of Numerical Analysis, we recommend pairing this book with such a textbook. See, for example, [2; 7; 8].

If you are still wondering whether to use a programming language such as Python™, C, C++, Java™, or some computational software such as Maple™, Mathematica®, Octave, or R, look no further. We strongly recommend you Sage Math. It is FREE, open-source, and it uses a Python-like syntax. This last phrase contains some of the strongest arguments why you may want to choose Sage Math.

To elaborate on the last statement, free open-source software allows you to use the software for free and also allows you to access the source code, which can be a great source of inspiration and information. It also allows you to obtain the entire Sage Math's source code and change it to better address your own needs.

The Sage Math syntax is similar to that of Python. Python has become very popular for being one of the easiest introductory programming languages. It became so trendy that many (if not most, by the time you read this book) universities in the United States are using Python to teach Introductory to Programming courses. It is that easy! In fact, most of the Sage Math source code was written in Python. An interesting article one may want to read in this direction is "Python bumps off Java as top learning language," by Joab Jackson (http://www.javaworld.com/article/2452940/learn-java /python-bumps-off-java-as-top-learning-language.html). It says that "Python has surpassed Java as the top language used to introduce U.S. students to programming and computer science."

We hope we got your interest in learning Sage Math. You can use it as a Computer Algebra System, as a Programming Tool, or to create nice interactive mathematical demonstrations (using Sage Interacts). You can download and install Sage Math on your own machine, or you can use it over the Internet: for this, you may either choose Sage Cloud, or you may prefer Sage Cell. Sage is accessible from your desktop computer or from your smartphone.

We recommend this book to all undergraduate Mathematics, Mathematics Education, Computer Science and Information Technology, pre-Engineering, and Science students and instructors. Sage Math can be used in most Mathematics courses, in Introduction to Programming courses, as well as other computational courses.

RAZVAN A. MEZEI
USA

1

INTRODUCTION

1.1 WHAT IS SAGE MATH?

If you got a copy of this book, you probably already know that Sage Math is a free open-source mathematics software that is a great alternative to other software such as Mathematica, Maple, Matlab, and even the TI-83/TI-84 calculators. Once you get to master Sage Math, you won't want to use anything else. It's a great tool, easy to use, and very intuitive. And if you can't find a specific function that you may need for a project, then you can easily program it yourself. You will learn how to do this as you read this book.

The official website for Sage Math is `http://Sagemath.org` [16]. On this website, you can find `Quickstart Manuals`, `Official Documentation Manuals`, and `Official Binaries` that you can use in order to install Sage Math on your own machine. Although the website is very nicely organized, you cannot overestimate the use of the `"Search"`, button which is also available. The `source code` is obtainable there too.

Note: As of February 2015, Sage announced that it will add "Math" to its title in order to disambiguate with other "Sages". Throughout this book, we will use both terms "Sage" and "Sage Math" interchangeably.

An Introduction to SAGE Programming: With Applications to SAGE Interacts for Numerical Methods,
First Edition. Razvan A. Mezei
© 2016 John Wiley & Sons, Inc. Published 2016 by John Wiley & Sons, Inc.

1.2 VARIOUS FLAVORS OF SAGE MATH

1.2.1 Sage Math on Your Machine

In order to use Sage Math, you can **install it** on your own computer. This way you won't need Internet connection to run Sage applications and you can also save your own work.

You can find the binaries on the official Sage Math website [16]. One can download binaries for Linux, Mac OS X, and Oracle Solaris. At the moment, the Windows machines need to install a Virtual Machine in order to use Sage Math on such systems. Detailed installation steps can be found here: `http://wiki.sagemath .org/SageAppliance`.

One can also download and use a **Live CD** with Sage.

1.2.2 Sage Cell

The author's favorite way to use Sage Math is through a **Sage Cell**. Using a web browser, one can run Sage Math without the need to install anything on their computer. Moreover, you won't need to worry about having the latest version of Sage Math installed on your computer. One such Sage Math Cell can be found here [13]. All the examples in this book were tested using this Sage Math Cell, running the following version: `'Sage Version 6.3, Release Date: 2014-08-10'`. Before you start using it, be aware of the following two main limitations: you need to be connected to the Internet, in order to use Sage Cell, and you won't be able to save your work in there. On the positive side it is very easy to use. It works well on desktop/laptop computers as well as on smartphones.

Note: We recommend you to try different browsers and see which one works the best with the Sage Cell you are using. It is the author's experience that some browsers will work significantly faster with the above-mentioned Sage Cell, than others.

1.2.3 Sage Cloud

Another flavor of Sage Math is using a **Sage Notebook** (`http://sagenb.org`). As the front page of this website mentions, one can use it to "create, collaborate on, and publish interactive worksheets". Once you register and create a free account, you can create Sage code, save it, access it, and even share it.

The latest development, the collaborative web-based interface of Sage Math is the **Sage Cloud** (`https://cloud.sagemath.com`), which seems to quickly replace Sage Notebook. It adds features and capabilities such as "collaboratively work with Sage Worksheets, IPython notebooks, LaTeX documents", Course Management (an example is a UCLA 400+ student Calculus course), and many others. There is even a Chrome App available that works with it. Sage Cloud is planned to replace Sage Notebook. One can even run code written in other programming languages such as C, C++, Java, and many others, inside Sage Cloud.

To create a free account, just follow the link posted on Sage Math main page (or go to `https://cloud.sagemath.com/`). There you will be invited to either sign in or create a free account.

To create Sage code and run it inside the Sage Cloud, you will first need to create a project. If you click on `"Create New Project"` button, you will be invited to select a name and an optional description. Then, clicking on the link `"Create or Import a File, Worksheet ..."`, one can select `Sage Worksheet`, and create a new Sage Worksheet. There, one can type in Sage code and run it.

Note: All the Sage Math code given in this book was tested using Sage Cell. As such, some of the code may need to be changed/tweaked in order to run in Sage Cloud.

For example, the following code runs well in the Sage Cell, but needs some tweaking for Sage Cloud:

```
#Here come the "fancy" Interacts
@interact
def myInteract1(
    f = input_box(default=e^x ),
    n = slider(vmin=0, vmax=10, step_size=1, \
               default=3, label="Select the order n: "),
    x0 = input_box(default=0 ),
    simplified = selector(values = ["Yes", "No"], \
               label = "Simplify: ",default = "No" )):
    if(simplified == "Yes"):
        print f, " = "  , f.taylor(x, x0, n).full_simplify()
    else:
        print f, " = "  , f.taylor(x, x0, n)
```

The following is a tweaked version of the previous code that runs on both Sage Cloud and Sage Cell:

```
@interact
def myInteract2(
    f = input_box(default=e^x ),  \
    n = slider( 0, 10, step_size=1, \
               default=3, label="Select the order n: "), \
    x0 = input_box(default=0 ),
    simplified = selector(values = ["Yes", "No"], \
               label = "Simplify: ",default = "No" )):
    if(simplified == "Yes"):
        print f, " = "  , f.taylor(x, x0, n).full_simplify()
    else:
        print f, " = "  , f.taylor(x, x0, n)
```

The webpage `https://github.com/sagemath/cloud/wiki/Teaching` contains a list of links to several courses (such as Calculus, Combinatorics, Statistical Computing, Cryptography, Computer Systems Security, Experimental Gravitational Wave Physics, Linear Algebra, Differential Equations, Abstract Algebra, and many others) that are using Sage Math See also: [1], [2], and [3].

Some great references that motivated this work are: [4, 6, 11, 12, 15].

2

USING SAGE MATH AS A CALCULATOR

Sage Math can easily be used as a calculator, one that has lots of "features." My favorite one is the fact that you can program your own algorithms in a programming language that is very easy to learn, but we'll learn this in the next chapter. In this chapter, we get introduced on how to use basic arithmetic expressions, as well as some Sage Math library functions that are already available to the user. We focus mostly on functions that are useful in a Numerical Analysis course such as solving equations, taking derivative and antiderivatives of a function, and finding the Taylor polynomial of degree n of a given function. Then we describe different functions and options that can be used to easily and efficiently obtain 2D and 3D plots.

Throughout the next two sections, we refer to the Sage Math code by including the optional (for the Sage Cell) word "`sage:`" in each line of code. You may choose to either type it into the Sage editor, or simply ignore it. The results will be the same. To run the Sage Math code given below, you should open your favorite "flavor" of Sage Math and type in the code as indicated. As mentioned earlier, all these examples were tested using Sage Cell (`'Sage Version 6.3, Release Date: 2014-08-10'`).

2.1 FIRST SAGE MATH EXAMPLES

We start with the typical first example in a programming course, the "Hello World!". First, type the following Sage code into the Sage Math editor:

An Introduction to SAGE Programming: With Applications to SAGE Interacts for Numerical Methods,
First Edition. Razvan A. Mezei
© 2016 John Wiley & Sons, Inc. Published 2016 by John Wiley & Sons, Inc.

```
sage: print "Hello World!"
```

or equivalently you could simply type only:

```
print "Hello World!"
```

and then click on the $\boxed{\textbf{Evaluate}}$ button (or press the following combination of keys: "shift + enter") to run the code. You will get (in either case) the following output:

```
Hello World!
```

To check which version of Sage you are currently using, run the following Sage code:

```
sage: version()
```

At the time of writing this book, we got the following:

```
'Sage Version 6.3, Release Date: 2014-08-10'
```

You will probably get a higher version.

2.2   COMPUTATIONS

2.2.1   Basic Arithmetic Operators

As you may expect, Sage Math has the following arithmetic operators: "+" for *addition*, "−" for *subtraction,* "/" for *division,* "*" for *multiplication*. As opposed to many programming languages (but similar to Python), Sage also has an *exponential operator*: "**". Besides it, Sage Math offers another *exponential operator*, albeit equivalent: "^". We mostly use "^" throughout this book, rather than "**".

The following examples should be pretty self-explanatory:

```
sage: 6+4
```

```
10
```

```
sage: 6-4
```

```
2
```

```
sage: 6*4
```

```
24
```

```
sage: 6/4
```

> 3/2

Note: When you divide two integers, you will neither get a floating-point number (as in Python) nor another integer (as in Java, C, and C++), but rather a simplified fraction. By default, Sage will try to give you an exact answer and this implies that Sage will merely simplify your expression as much as it can. One "trick" to obtain a decimal value is by making at least one of the operands a floating-point (a decimal) number. For example,

```
sage: 6/4.0
```

> 1.50000000000000

```
sage: 6**4
```

> 1296

```
sage: 6^4
```

> 1296

```
sage: --4
```

> 4

Note: As opposed to Java and C/C++ (and many other programming languages) where division of integer values yields an integer result, the division of two integers in Sage may be a fraction. Sage Math will actually give you an exact, reduced, fraction. By default, Sage Math will not give you a decimal representation of the answer. So, 2/4 will not yield 0 or 0.5, but 1/2. If you need to get an integer value when you divide two integers (and remove the fractional part), then you can use the following operator: "//". This is similar to Python. For example, 6//4 will output 1.

```
sage: 6//4
```

> 1

To get the remainder in a division, one needs to use the "%" operator (also called the *modulus* operator).

```
sage: 6%4
```

> 2

Application: How many days are in 2015 hours? If now is 7 pm, what time will it be 2015 hours later?

Answer: To use Sage Math, we can use the following expression: 2015//24. Hence

sage: 2015//24

```
83
```

which means there are 83 days in 2015 hours. For the second part, one needs to compute the remainder of 2015 divided by 24. Hence

sage: 2015%24

```
23
```

means that 2015 hours later it will be 6 pm.

The order of operations is the same as we would expect from any calculator: first come parentheses, then exponents, multiplication, division, addition, and subtraction. If we wish to give any operation a higher precedence, we should simply put it in parentheses.

sage: 6+4^2

```
22
```

which is equivalent to

sage: 6+(4^2)

and one can see that the exponential operator was applied before the addition. If we want the addition to be performed before exponential, we can use parentheses as in the following example:

sage: (6+4)^2

```
100
```

Note: For nested parentheses one cannot use brackets ("[" and "]") or curly braces ("{" and "}"), since these are reserved for other purposes in Sage Math. So an expression such as

$$\{[(2-4)*(4-3)+4]\hat{}10\}/(2\hat{}20)$$

can be evaluated using the following Sage code:

sage: (((2-4)*(4-3)+4)^10)/(2^20)

and obtain

```
1/1024
```

Using parentheses, one can modify the order of operations, since parentheses have the highest precedence:

```
sage: 2+2*2-2 , 2+(2*2)-2 ,  (2+2)*2-2 ,  (2+2)*(2-2)
```

will yield

```
(4, 4, 6, 0)
```

Note: Above, we included four arithmetic expressions separated by commas. When we want Sage to evaluate several expressions in the same line, we simply write each of these expressions and then separate them by commas.

Sage Math can easily handle large numbers:

```
sage: 2.0^10000
```

gives

```
1.99506311688076e3010
```

and

```
sage: 2^10000
```

yields (very quickly!)

19950631168807583848837421626835850838234968318861924548520089498529438830221946631919961684036194597899331129
42320912427155649134941378111759378593209632395785573004679379452676524655126605989552055008691819331154250860
84606181046855090748660896248880904898948380092539416332578506215683094739025569123880652250966438744410467598
71626985453222868538161694315775629640762836880760732228535091641476183956381458969463899410840960536267821064
62142733339403652556564953060314268023496940033593431665145929777327966577560617258203140799419817960737824568
37622800373028854872519008344645814546505579296014148339216157345881392570953797691192778008269577356744441230
62018757836325502728323789270710373802866393031428133241401624195671690574061419654342324638801248856147305207
43199225961179625013099286024170834080760593232016126849228849625584131284406153673895148711425631511108974551
42033138202029316409575964647560104058458415660720449628670165150619206310041864222759086709005746064178569519
11456055068251250406007519842261898059237118054447880729063952425483392219827074044731623767608466130337778706
03980341319713349365462270056316993745550824178097281098329131440357187752476850985727693792643322159939987688
66080836883783802764328277517227365757274478411229438973381086160742325329197481312019760417828196569747589811
64531258434135959862784130128185406283476649088690521047580882615823961985770122407044330583075869039319604603
40497315658320867210591330090375282341553974539439771525745529051021231094732161075347482574077527398634829849
83407569379556466386218745694992790165721037013644331358172143117913982229838458473344402709641828510050729277
48364550578634501100852987812389473928699540834346158807043959118985815145779177143619698728131459483783202081
47498217185801138907122825090582681743622057747592141765371568772561490458290499246102863008153558330813010198

76758562343435389554091756234008448875261626435686488335194637203772932400944562469232543504006780272738377553
76406726898636241037491410966718557050759098100246789880178271925953381282421954028302759408448955014676668389
69799688624163631337639390337345580140763674187771105538422573949911018646821969658165148513049422236994771476
30691554682176828762003627772577237813653316111968112807926694818872012986436607685516398605346022978715575179
47385246369446923087894265948217008051120322365496288169035739121368338393591756418733850510970271613915439590
99159815465441733631165693603112224993796999922678173235802311186264457529913575817500819983923628461524988108
89602322443621737716180863570154684840586223297928538756234865564405369626220189635710288123615675125433383032
70029097668650568557157505516727518899194129711337690149916181315171544007728650573189557450920330185304847113
81831540732405331903846208403642176370391155063978900074285367219628090347797453332046836879586858023795221862
91200807428195513179481576244482985184615097048880272747215746881315947504097321150804981904558034168269497871
41316063210686391511681774304792596709376

2.2.2 Decimals Versus Exact Values

As mentioned earlier, if an arithmetic expression involves only integers, then Sage Math will try to return an exact simplified answer.

```
sage: 6/4
```

> 3/2

If the expression, however, involves at least one number that is using the floating-point representation, then Sage will return a decimal, floating-point representation of the answer. **This however may not always be exact**!

```
sage: 6/4.0
```

> 1.50000000000000

If you have an expression that contains only integers, but you want a decimal representation of the result, you can use the following:

```
sage: 6/4.n()
```

> 1.50000000000000

By default, the decimal representation uses 53 bits precision. If you need more (or less) precision, you can specify it as in the following example:

```
sage: 6/4.n(53)
```

> 1.50000000000000

```
sage: 6/4.n(10)
```

> 1.5

```
sage: 6/4.n(100)
```

```
1.5000000000000000000000000000000
```

Note: Please note that a 10-bit precision is about three decimal digits precision (since 2^10 is 1024).

Note: The following expressions yield the same results: `(6.4).n(53)`, `(6.4).n()`, and `n(6.4)`.

Note: Perhaps easier to use is the `round` method, which accepts a second parameter, the number of *decimal* places. For example,

```
sage: round(pi, 3), round(pi, 5)
```

```
(3.142, 3.14159)
```

```
round(1/4, 2), round(1/4, 5), round(1/4, 1)
```

```
(0.25, 0.25, 0.3)
```

At the moment of writing this book, we find `round` as a useful but limited method. For example, the following Sage Math expressions unexpectedly produce the same output:

```
sage: round(pi, 15), round(pi , 50)
```

```
(3.141592653589793, 3.141592653589793)
```

2.2.3 Constants

Sage has a set of built-in symbolic constants. Examples of them are π (pi), ∞ (oo – which is two lowercase letters of "o"), e (e), i (can be either "i" or "I"), ϕ (golden_ratio), euler_gamma, and so on.

If one wishes to print several decimal places of the value of π, then the following Sage code can be handy:

```
sage: pi.n()
```

```
3.14159265358979
```

To compute $e^{\pi i}$ one can easily use the following expression:

```
sage: e^(pi*i)
```

```
-1
```

Note: Although Sage Math is *case-sensitive* (which means the case of the characters matters!), we can use either i or I to represent $\sqrt{-1}$.

```
sage: i^2
```

> ```
> -1
> ```

```
sage: I^2
```

> ```
> -1
> ```

Another well-known constant is the golden ratio $\frac{1+\sqrt{5}}{2}$.

```
golden_ratio.n()
```

> ```
> 1.61803398874989
> ```

The following will be used later in this book: ∞. To enter it, you can either use

```
infinity
```

or

```
Infinity
```

or two little oh letters

```
oo
```

To obtain $-\infty$, simply put a negative sign before ∞.

2.2.4   Breaking Long Lines of Code

Similarly to Python, whenever you have a line of code that is too long and you want to break it into multiple lines, you can use "\" to split it. Make sure you don't use it inside a **string** (strings are discussed later in the book) or a number.

For example,

```
sage: (((2-4)*(4-3)+4)^10)/(2^20)
```

can be split as follows:

```
sage: (((2-4)*(4-3)+   \
sage: 4)^10)/(2^20)
```

to obtain

> ```
> 1/1024
> ```

or equivalently

```
sage: (((2-4)*      \
sage: (4-3)+        \
sage: 4)^10)/(2^20)
```

We use this feature quite extensively throughout this book.

Very important note: The *indentation* (the spaces at the beginning) of each line of code is very important. Consecutive lines of code that have the same indentation are part of the same block of statements. We'll give more details on this later in the book, but for now, it is important to note that the indentation is very important and therefore be careful not to indent unnecessarily these lines. This is different than C++/Java, where white spaces are ignored by the compiler and the statements are grouped using { and }.

2.2.5 Comments

To help make the code easier to read, one should use *comments*. A **comment** is the part of the code that starts with the symbol "#" (just as in Python) and goes all the way until the end of that line. Comments are completely ignored by Sage, but they do help the reader follow the code more easily.

Note: Many beginning programmers have a tendency of disliking to include comments in their code, but as they progress and get involved in larger programs, they get to really appreciate having good meaningful comments throughout the code.

Here is an example of how comments can be used effectively to make the code easier to read and understand. Let's solve the following problem:

A loan of $7000 was repaid at the end of 7 months. What size repayment check (principal and interest) was written, if an 8% annual rate of interest was charged?

We give the following solution. The explanation is given using comments

```
sage: #using the simple interest formula
sage: #the future value is S = P(1+r*t)
sage: #P = invested amount
sage: #r = annual interest rate = 0.08
sage: #t = time measured in years = 7/12 years
sage: 7000*(1+0.08*7/12)
```

We obtain

```
7326.66666666667
```

We'll learn later in this book how to format our output in a more user-friendly way. For now, you can simply use

```
sage: (7000*(1+0.08*7/12)).n(24)
```

to obtain

```
7326.67
```

One could also use

```
sage: round((7000*(1+0.08*7/12)), 2)
```

and get the same output as above.

Note: Multiline comments are also available in Sage. They are enclosed in " '. They are the equivalents of the /* */ used in C++ and Java programming languages. For example,

```
"'this is a multiline
comment"'
```

is a comment that spans across two lines and it will be ignored by Sage.

2.2.6   Library Functions

We have seen how one can easily evaluate arithmetical expressions using Sage. But Sage Math can do so much more! Next we'll see a very small set library functions that come with Sage. This is the part where the power of Sage starts to be more appealing to the user.

Note: From here on we'll drop the optional "sage:" throughout the remaining of the book.

In these examples, we mostly give the explanation of the code embedded as comments.

```
#to compute the square root of 72
#simply use the sqrt function
sqrt(72)
```

```
6*sqrt(2)
```

```
#to compute a decimal representation
#of square root of 72 use the n()
sqrt(72).n()
```

```
8.48528137423857
```

```
#check whether a number is prime
(2^50-1).is_prime()
```

```
False
```

```
#to factor an integer expression
(2^50-1).factor()
```

```
3 * 11 * 31 * 251 * 601 * 1801 * 4051
```

```
#to factor an integer expression (2)
(10000).factor()
```

```
2^4*5^4
```

```
#compute the greatest common divisor
# of two integers
gcd(6,4)
```

```
2
```

```
#compute the least common multiple
# of two integers
lcm(6,4)
```

```
12
```

```
#absolute value
abs(-2015)
```

```
2015
```

```
#natural logarithm
ln(e^2)
```

```
2
```

```
#NOTE: log also denotes the natural logarithm
#      that is log base e
# it does not denote the common logarithm
#      which has base 10
log(e^2)
```

```
2
```

```
#NOTE: another log example
log(10.0)
```

```
2.30258509299405
```

```
#NOTE: log base 2
log(32, 2)
```

```
5
```

```
#to obtain log base 10 use
log(1000, 10)
```

```
3
```

```
#NOTE: log also denotes natural logarithm
log(1024.0, 10)
```

> 3.01029995663981

```
#trigonometric functions are available too
sin(pi/4)
```

> 1/2*sqrt(2)

```
#more on trigonometric functions
(  sin(pi/9)+cos(0)-tan(pi/4)  ).n()
```

> 0.342020143325669

```
#find a list of all divisors of a number
(123^4-7).divisors()
```

> [1, 2, 114443317, 228886634]

```
#compute the inverse of a matrix
matrix([[1,2], [3,2]])^(-1)

#OR equivalently
A = matrix([[1,2], [3,2]])
A^(-1)
```

> [-1/2 1/2]
> [3/4 -1/4]

```
#compute the determinant of a matrix
A = matrix([[1,2], [3,2]])
det( A )
```

> -4

```
#compute the power of a matrix
matrix([[1,2], [3,2]])^(10)
```

> [419431 419430]
> [629145 629146]

```
#compute the product of two matrices
A = matrix([[1,2], [3,2], [5,6]])
B = matrix([[1,2,6,7], [3,2,6,7]])
A*B
```

> [7 6 18 21]
> [9 10 30 35]
> [23 22 66 77]

```
#to compute a sum
#let's say the sum of the first 10 squared integers
sum([x^2 for x in [1,2,..,10]])
#OR equivalently:
sum([x^2 for x in [1..10]])
```

385

Note: One can use the TAB key to get Sage autocomplete the name of a function. This comes very handy, especially when you don't know the exact spelling of a function, or you want to check if a particular function already exists. For example, type `is_` and then press the TAB key. You should obtain a list of available functions that start with `is_`. In this list, one can note functions such as `is_even`, `is_power_of_two`, `is_prime`, `is_pseudoprime`, and `is_squarefree`.

2.2.7 Working with Strings

To output a value or a text, one can use the `print` statement. This function is especially useful if we need to output to the user multiple lines.

Strings (see also p. 80) are text that are included between single or double quotes. When we output a string, Sage Math will print the text **as is**. If we remove the quotes, then Sage will attempt to *evaluate* it.

For example,

```
print '4+6'
```

will output

4+6

```
print 4+6
```

will output

10

For a nice, user-friendly output, one should use both strings and values. The previous can be combined as

```
print '4+6 = ', 4+6
```

to output

4+6 = 10

Note: To print multiple expressions in the same line, one can either use commas to *separate* them or use "+" to *combine* them into one string. "Adding" two strings will result in concatenating them into one larger string.

```
print 'this is' + "how to add"+"2 strings"
```

will output (note the missing space between is and how!):

```
this ishow to add2 strings
```

Note: You cannot directly "add" (append) a number to a string. You would first have to convert the number into a string using the str function.

```
#print   '4+6 = '+ 4+6    !!this won't work!!
#    the next line will work!
print    '4+6 = ' + str(4+6)
```

will output

```
4+6 = 10
```

If you want to include *single* (similarly for *double*) *quotation marks* in your string, you simply have to use the double quotation marks to delimitate the string. For example,

```
print "O'Brian's"
```

yields

```
O'Brian's
```

One could also use *escape sequences* such as \ ', \ ", and \ \ in order to output ', ", and \. For example,

```
print "This is a single quote: \', and this is a backslash: \\"
```

outputs

```
This is a single quote: ' and this is a backslash: \
```

Other escape sequences that we use in this book inside a string are
\ t for tab
\ n for newline
There are many other important information to say about strings but we will mention them in the next chapter.

2.2.8 Solving Equations and Inequalities

One can use the `solve` method to solve an equation or a system of equations. For example,

```
#solve(equation, variable to solve for)
solve(x^2+x==8,x)
```

will output

```
[x == -1/2*sqrt(33) - 1/2, x == 1/2*sqrt(33) - 1/2]
```

```
solve ( exp ( x ) == -1/2 , x )
```

will output

```
[x == I*pi + log(1/2)]
```

To solve a system of equations, we need to declare the variables being used.

```
#declare the variables that will be used.
#          x is by declared by default!
var('y')
#solve([equations], variables to solve for)
solve ([x+y==8, x-2*y==0] , x, y )
```

will output

```
[[x == (16/3), y == (8/3)]]
```

The next system has infinitely many solutions:

```
#declare the variables that will be used
#   we can either use var('y,z') or we can use:
var('y z')
#then solve([equations], variables to solve for)
solve ([x+y==8, x-2*z==0] , x, y,z )
```

will output

```
[[x == r1, y == -r1 + 8, z == 1/2*r1]]
```

The following system has no solutions:

```
#declare the variables use. x is by default declared
var('y z')
#then solve([equations], variables to solve for)
solve ([x+y==8, x-2*z==0, x-2*z==1] , x, y,z )
```

will output

```
[]
```

Note: In Sage Math (just as in Java, C/C++, C#, Python, and many other programming languages), the *assignment* operator is "=", while the *comparison* operator is "==". Hence, whenever you have an equation, you should use "==".

If Sage cannot solve the equation exactly, as in the following example:

```
#Sage did not find the exact solution
solve(x^5+x==8,x)
```

> [0 == x^5 + x - 8]

then you could use the find_root function, which requires an interval where to search for the solution. It is important to know that even if more than one solution is found, this method will only return one of them.

```
find_root(x^5+x==8, -2, 2)
```

> 1.456023377174738

The same solve function (as seen above) can be used to solve for inequalities:

```
solve(x^2-6>=8,x)
```

> [[x <= -sqrt(14)], [x >= sqrt(14)]]

```
solve(x^2-6<0,x)
```

> [[x > -sqrt(6), x < sqrt(6)]]

Note how solutions are enclosed in individual brackets as in [[x <= -sqrt (14)], [x >= sqrt(14)]]. This means that we have two distinct solutions: $(-\infty, -\sqrt{14}]$ and $[\sqrt{14}, \infty)$.

In [[x > -sqrt(6), x < sqrt(6)]], we should read that the solution is the interval $(-\sqrt{6}, \sqrt{6})$.

Note: As stated above, whenever we are using several variables, we need to declare them. By default, x is already declared; hence, we only need to declare the other variables (albeit 'x' can be declared again).

To solve the following system of inequalities:

$$\begin{cases} x+y \geq 10 \\ x-y < 20 \end{cases}$$

one can use the following Sage code:

```
var('y')
solve ([ x + y >=10 , x - y < 20] , x , y )
```

and obtain

> [[x == -y + 10, -5 < y], [-y + 10 < x, x < y + 20, -5 < y]]

2.2.9 Calculus Functions

To define a mathematical function and use it in computations, one can use code similar to the one below:

```
f(x) = cos(x)^7
```

Then, you can use Sage Math functions to evaluate it at specific values or compute its derivative, antiderivative, Taylor series, and so on.

```
#define the function f
f(x) = cos(x)^7
#then evaluate f at x = pi
f(pi)
```

and obtain

```
-1
```

```
#define the function f
f(x) = cos(x)^7
#then evaluate f at x = pi
f(pi/6)
```

will output

```
27/128*sqrt(3)
```

Using the show function, one can get a much nicer output:

```
#define the function f
f(x) = cos(x)^7
#then evaluate f at x = pi
show(  f(pi/6)  )
```

$$\frac{27}{128}\sqrt{3}$$

If one needs to obtain a decimal value, then the following Sage code

```
#define the function f
f(x) = cos(x)^7
#evaluate f at x = pi
f(pi/6).n()
```

will output

```
0.365354467221560
```

To find the derivative of this function f, one can use

```
f(x) = cos(x)^7
f(x).diff()
```

or

```
f(x) = cos(x)^7
f(x).derivative()
```

or

```
f(x) = cos(x)^7
derivative(f(x))
```

or

```
f(x) = cos(x)^7
diff(f(x))
```

or simply

```
diff(cos(x)^7)
```

to obtain

```
-7*cos(x)^6*sin(x)
```

For a nicer output of the same answer, one can use the show function

```
f(x) = cos(x)^7
show( f(x).diff()   )
```

and get

$$-7 \cos (x)^6 \sin(x)$$

Similar code can be used in order to compute the antiderivative of a function:

```
f(x) = cos(x)^7
f(x).integrate(x)
```

or simply

```
integrate(cos(x)^7, x)
```

will display

```
-1/7*sin(x)^7 + 3/5*sin(x)^5 - sin(x)^3 + sin(x)
```

To compute the definite integral of a function on a given interval, let's say $\left[0, \frac{\pi}{3}\right]$ one can use

```
f(x) = cos(x)^7
f(x).integrate(x, 0, pi/3)
```

or equivalently

```
f(x) = cos(x)^7
integrate(f(x), x, 0, pi/3)
```

or simply

```
integrate(cos(x)^7, x, 0, pi/3)
```

and get

```
1181/4480*sqrt(3)
```

To compute the second derivative of the function `f(x)`, you can use

```
f(x) = cos(x)^7
f(x).derivative(x, 2)
```

and obtain

```
-7*cos(x)^7 + 42*cos(x)^5*sin(x)^2
```

To find the fifth-order derivative of the function, one can use Sage code similar to

```
f(x) = cos(x)^7
f(x).derivative(x, 5)
```

To compute the Taylor polynomial of degree (at most) n, around the point c, use `taylor(x,c,n)` as below:

```
f(x)=cos(x)^7
#taylor polynomial of degree 7, around c=0
f(x).taylor(x, 0, 7)
```

to obtain

```
-3787/720*x^6 + 133/24*x^4 - 7/2*x^2 + 1
```

To compute the Taylor polynomial of degree 7 of $f(x) = e^x$ around $c = 1$, one can use

```
f(x)=e^x
f(x).taylor(x, 1, 7)
```

and get

```
1/5040*(x - 1)^7*e + 1/720*(x - 1)^6*e + 1/120*(x - 1)^5*e +
1/24*(x - 1)^4*e + 1/6*(x - 1)^3*e + 1/2*(x - 1)^2*e + (x - 1)*e + e
```

Note: To obtain a fully simplified answer, one can use the `full_simplify()` as below:

```
f(x)=e^x
f(x).taylor(x, 1, 7).full_simplify()
```

and obtain

```
1/5040*x^7*e + 1/240*x^5*e + 1/72*x^4*e + 1/16*x^3*e +
11/60*x^2*e + 53/144*x*e + 103/280*e
```

Remember to use `show()` for a nicer display:

```
f(x)=e^x
show( f(x).taylor(x, 1, 7).full_simplify() )
```

$$\frac{1}{5040}x^7e + \frac{1}{240}x^5e + \frac{1}{72}x^4e + \frac{1}{16}x^3e + \frac{11}{60}x^2e + \frac{53}{144}xe + \frac{103}{280}e$$

Let $f(x) = \frac{3\sin x}{x}$. From the way it is defined, we can easily see that $f(0)$ is undefined, but $\lim_{x\to 0} f(x) = 3$. In Sage Math, we can see this by trying to run the following Sage code:

```
f(x)=3*sin(x)/x
f(0)
```

which will return an error message ("`division by zero`"). Computing the limit,

```
f(x)=3*sin(x)/x
limit( f, x = 0)
```

we obtain

```
x |--> 3
```

One can define functions and then use them in some algebraic expressions. For example, the following example will compute the composition of two functions:

```
f(x) = x^2+7
g(x) = (x-2)
```

```
f(g(x))
```

which outputs the composition of the two functions

```
(x - 2)^2 + 7
```

If needed, one may use `full_simplify()`.

2.2.10 Exercises

(1) Compute a decimal representation of $\sqrt{7}$.

(2) Compute a decimal representation of $\sqrt[3]{14}$.

(3) Compute a decimal representation of $\sqrt{-7}$.

(4) Compute a decimal representation of $\sqrt[3]{-14}$.

(5) Compute the following expression using 10-bit decimal representation:

$$\frac{20 - 55}{40 - 50}.$$

(6) Compute the following expression using 10-bit decimal representation:

$$\frac{(20 - 55)^2}{(40 - 50)^3}.$$

(7) Compute the following expression using 10-bit decimal representation:

$$\left[(10 + \sqrt{2})^2\right]^4.$$

(8) Compute the following expression using 10-bit decimal representation:

$$\left[(10 - \sqrt[3]{5})^3\right]^4.$$

(9) How many days are there in 10,000 hours? If right now is 7:01 am, what time will it be 10,000 hours later?

(10) How many days are there in 12,345 hours? If right now is 9:01 pm, what time will it be 12,345 hours later?

(11) How many hours are there in 10,000 minutes? If right now is 7:01 am, what time will it be exactly 10,000 minutes later?

(12) How many hours are there in 12,345 minutes? If right now is 9:01 pm, what time will it be exactly 12,345 minutes later?

(13) Use Sage code to print at least 10 digits for the decimal representation of the golden ratio.

(14) Use Sage code to print at least 10 digits for the decimal representation of e.

(15) Determine whether $2^{2015} - 1$ is prime or not.

(16) Determine whether $4^{2015} - 1$ is prime or not.

(17) Factor $3^{123} - 1$.

(18) Factor $5^{123} - 1$.

(19) Find the lcm and gcd for the following pair of whole numbers: $2^{30} - 1, 2015$.

(20) Find the lcm and gcd for the following pair of whole numbers: $30^5 - 1, 2^{200}$.

(21) Find the list of all divisors of $2^{80} - 1$.

(22) Find the list of all divisors of $3^{50} - 7$.

(23) What is the result of each of the following expressions?
- `9//6`
- `9/6`
- `9/6.0`
- `9%6`

(24) What is the result of each of the following expressions?
- `12//8`
- `12/8`
- `12/8.0`
- `12%8`

(25) Find the number of digits of the following number: $2^{2^{100}}$. Hint: You may want to simplify your expression before using Sage.

(26) Find the number of digits of the following number: $3^{3^{100}}$. Hint: You may want to simplify your expression before using Sage.

(27) Find the number of digits of the following number: 2015^{2015}. Hint: You may want to simplify your expression before using Sage.

(28) Find the number of digits of the following number: $2015^{2014} - 2014^{2015}$. Hint: You may want to simplify your expression before using Sage.

(29) Correct the following Sage code so you can run it: `print 'cos(pi) = '+` `cos(pi)`.

(30) Correct the following Sage code so you can run it: `print 'ln(7) = '+` `ln(7)`.

(31) Solve the following equation: $3x^3 - 4x + 5 = 9$.

(32) Solve the following equation: $3x^2 - 4x + 5 = 19$.

(33) Find an approximate solution for the following equation: $3x^{25} - 4x + 5 = 19$. Note: You will need to find an interval first.

(34) Find an approximate solution for the following equation: $\cos(x) = \frac{\pi}{4}$.
Note: You will need to find an interval first.

(35) Find an approximate solution for the following equation: $\sin^2(x) = \frac{\pi}{4}$.
Note: You will need to find an interval first.

(36) Find an approximate solution for the following equation: $\sin(x) = e^x$.
Note: You will need to find an interval first.

(37) Find an approximate solution for the following equation:
$\sin(x) = \cos(x)$ in $\left[0, \frac{\pi}{2}\right]$.

(38) Solve the inequality: $x^2 - 4x + 7 > 10$.

(39) Solve the inequality: $x^2 - 4x + 7 \leq 10$.

(40) Solve the system of equations:

$$\begin{cases} x + y + z = 10 \\ x + 2y + 3z = 20 \\ x + 4y + 9z = 30 \end{cases}.$$

(41) Solve the system of equations:

$$\begin{cases} x + y + z = 100 \\ x + 2y + 3z = 200 \\ x + 4y + 9z = 300 \end{cases}.$$

(42) Solve the system of equations:

$$\begin{cases} x + y + z = 10 \\ x + 2y + 3z = 20 \end{cases}.$$

(43) Solve the system of equations:

$$\begin{cases} x + y + z = 100 \\ x + 2y + 3z = 200 \end{cases}.$$

(44) For $f(x) = (\cos(x) + \sin(x))^{10}$, find $f'(x), f''(x), f'''(x),$ and $f^{(4)}(x)$.

(45) For $f(x) = (x - e^x)^{10}$, find $f'(x), f''(x), f'''(x),$ and $f^{(4)}(x)$.

(46) For $f(x) = (\cos(x) + \sin(x))^{10}$, find $f^{(7)}(x)$.

(47) For $f(x) = (x - e^x)^{10}$, find $f^{(7)}(x)$.

(48) For $f(x) = (x - e^{2x}\cos(x))^5$, find $\frac{d}{dx}f(x)$ and $\frac{d}{dy}f(x)$.

(49) For $f(x) = e^{2x}\cos^5(x)$, find $\frac{d}{dx}f(x)$ and $\frac{d}{dy}f(x)$.

(50) Find $\int (\cos(x) + \sin(x))^{10}dx$.

(51) Find $\int (x - e^x)^{10}dx$.

(52) Find $\int_0^\pi (\cos(x) + \sin(x))^{10}dx$.

(53) Find $\int_0^\pi (x - e^x)^{10} dx$.

(54) Find the Taylor polynomial of degree 10 for $f(x) = e^x$, around $c = 0$.

(55) Find the Taylor polynomial of degree 10 for $f(x) = \cos(x)$, around $c = 0$.

(56) Find the Taylor polynomial of degree 10 for $f(x) = \ln(x)$, around $c = 1$.

(57) Find the Taylor polynomial of degree 10 for $f(x) = \sin(x)$, around $c = 0$.

(58) Find the Taylor polynomial of degree 10 for $f(x) = e^x$, around $c = \ln 2$.

(59) Find the Taylor polynomial of degree 10 for $f(x) = \cos(x)$, around $c = \pi$.

(60) Find the Taylor polynomial of degree 10 for $f(x) = \ln(x)$, around $c = e$.

(61) Find the Taylor polynomial of degree 10 for $f(x) = \sin(x)$, around $c = \pi$.

(62) Let $A = \begin{pmatrix} 1 & -2 & 3 \\ 1 & 0 & 1 \\ 1 & 2 & 0 \end{pmatrix}$. Compute A^{20}.

(63) Let $A = \begin{pmatrix} 1 & 3 & 0 \\ 1 & 0 & 1 \\ 1 & 2 & 0 \end{pmatrix}$. Compute A^{10}.

(64) Approximate the solutions of the equation

$$\sin(x) = x^2.$$

(65) Approximate the solutions of the equation

$$\cos(x) = x^3.$$

(66) Approximate the solutions of the equation

$$\sin(x)\cos(x) = x^2 + 4x + 1.$$

(67) Approximate the solutions of the equation

$$\frac{x}{\sqrt{x^2 + 1}} = x^4 - x.$$

2.3   GRAPHS

In this section, we see how Sage Math can produce the graph of a given function (in both 2D and 3D), and plot a given set of points. This feature is especially important in Numerical Analysis and we'll use it throughout the remaining of the book.

2.3.1   2D Graphs

To plot the function $f(x) = \sin(x)$ on the interval $[-2\pi, 2\pi]$, one can use Sage code similar to the one below:

```
plot(sin(x), x, -2*pi, 2*pi)
```

or equivalently

```
p = plot(sin(x), x, -2*pi, 2*pi)
p.show()
```

and obtain

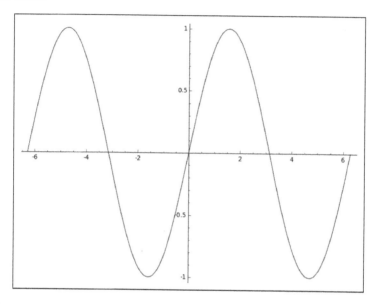

The previous code can be rewritten equivalently as

```
f(x) = sin(x)
a = -2*pi
b = 2*pi
p = plot(f(x), x, a, b)

p.show()
```

If you want to create a Portable Network Graphics (png) file and save it to your computer, rather than display the graph, then use the following code

```
p = plot(sin(x), x, -2*pi, 2*pi)
p.save('name.png')
```

Sage Math will give you a link to the temporarily created png file. In our example, we chose to name the file "name.png".

name.png [Updated 0 time(s)]

If you simply change the .png into .pdf you will obtain a Portable Document Format (pdf) file. Other allowed file extensions are " .eps ", " .ps ", " .sobj ", and " .svg ".

Next we'll see other useful features of the `plot2D`. Suppose we don't like the default `'blue'` color of the graph. To draw the graph using 'red' color, one needs to add in `plot2D` the following option: color='red' namely

```
p = plot(sin(x), x, -2*pi, 2*pi, color = 'red')
p.show()
```

and obtain

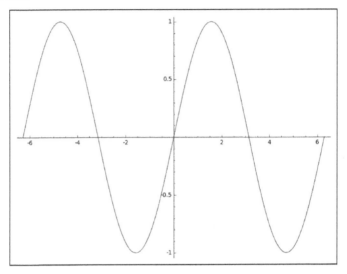

To use a different linestyle, one can use an option such as "`linestyle='--'`"

```
p = plot(sin(x), x, -2*pi, 2*pi, linestyle='--')
p.show()
```

and get

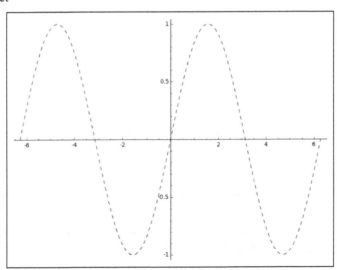

Other possible linestyles (see [16]) are

- "−" or "solid"
- "–" or "dashed"
- "−." or "dashdot"
- ":" or "dotted"

We recommend you to try them all as an exercise.
To the previous example, we next add a thickness thickness = 3:

```
p = plot(sin(x), x, -2*pi, 2*pi, linestyle='--', thickness = 3)
p.show()
```

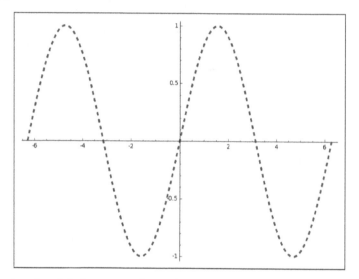

Changing the thickness option to 9 we get

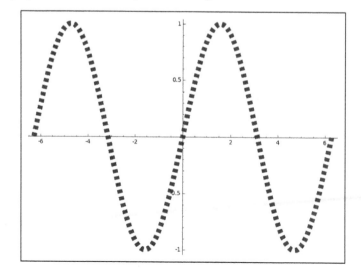

and for `thickness = 99`

To label the axes, one can include the following options: `axes_labels = [ 'x', 'sin(x)' ].`

```
p = plot(sin(x), x, -2*pi, 2*pi, color = 'purple', \
                    axes_labels =[ 'x','sin(x)'])
p.show()
```

produces

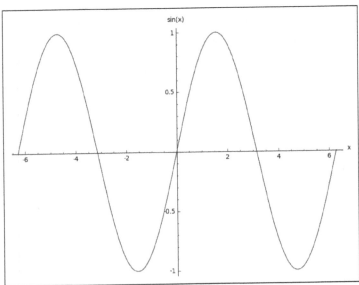

If you want the axes to use values such as $\frac{\pi}{3}, \frac{2\pi}{3}, \ldots,$ one can use the following options: `ticks = pi/3, tick_formatter=pi`:

```
p = plot(sin(x), x, -2*pi, 2*pi,                \
            color = 'purple',                    \
            axes_labels =[ 'x','sin(x)'],        \
            ticks = pi/3,                        \
            tick_formatter=pi)
p.show()
```

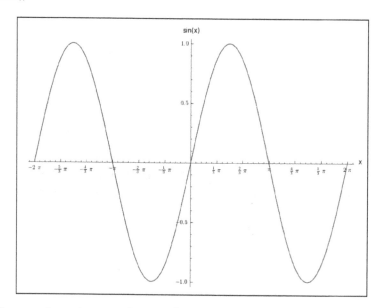

If one needs to plot more than one function in the same graph, then code similar to the next one could be used:

```
p = plot(sin(x), x, -2*pi, 2*pi,                \
            color = 'blue',                      \
            axes_labels =[ 'x','sin(x)'],        \
            ticks = pi/3,                        \
            tick_formatter=pi)

q = plot(sin(6*x), x, -2*pi, 2*pi,              \
            color = 'red',                       \
            linestyle = 'dotted')
(p+q).show()
```

and get

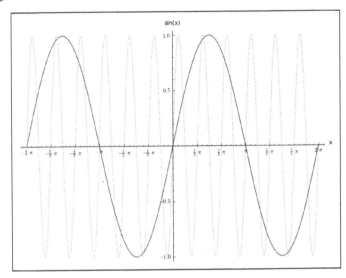

In the previous graph, one can also include a legend by using the following option: `legend_label`. The label can use strings or TeX commands as seen below:

```
p = plot(sin(x), x, -2*pi, 2*pi, color = 'blue',  \
         axes_labels =[ 'x','sin(x)'],            \
         ticks = pi/3, tick_formatter=pi,         \
         legend_label = 'sin(x)' )
q = plot(sin(6*x), x, -2*pi, 2*pi, color = 'red', \
         linestyle = 'dotted',                    \
         legend_label = '$sin(6 \cdot x)$' )
(p+q).show()
```

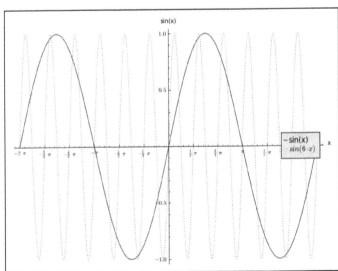

To change the *window range* of the graph, one can use the following options in the show method: `xmin=-2, xmax=2, ymin=-1/2, ymax=1/2`. So changing the last line of code in the previous example into

```
(p+q).show(xmin=-2, xmax=2, ymin=-1/2, ymax=1/2)
```

will generate

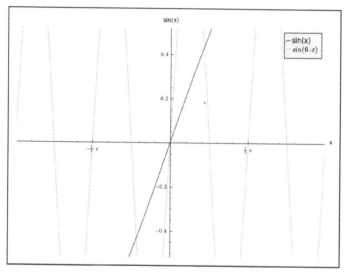

To plot two graphs on a separate set of axes, one can use Sage code similar to

```
graph1 = plot(sin(x), (x, -4*pi, 4*pi))
graph2 = plot(sin(6*x), (x, -4*pi, 4*pi))
show(graphics_array([graph1, graph2], 2, 1))
```

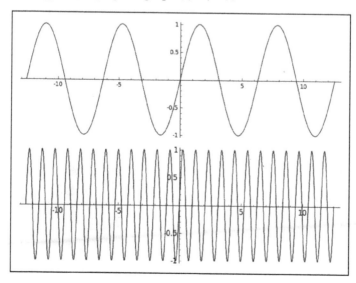

Changing the last line into

```
show(graphics_array([graph1, graph2], 1, 2))
```

we obtain

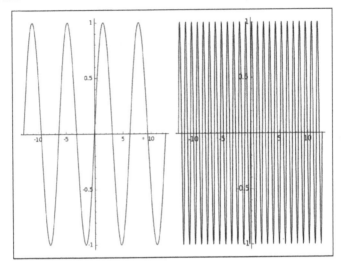

Instead of the 1×2 or 2×1 matrix used above, one could use other sizes. For example,

```
graph1 = plot(sin(x), (x, -4*pi, 4*pi))
graph2 = plot(sin(6*x), (x, -4*pi, 4*pi))
graph3 = plot(sin(12*x), (x, -4*pi, 4*pi))
empty  = plot(x,x, -1,1)
show(graphics_array(                    \
    [[graph1, graph2,empty],            \
     [empty ,empty, graph3]],    2, 3))
```

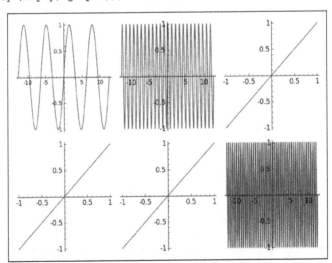

One can also include **text** in a graph, **plot a single point**, and **draw lines**. We'll see this through an example.

Let $f(x) = e^x$. Draw the graph of the function f in the interval $[-0.5, 3.5]$. Then draw the secant line that connects the points $(1, f(1))$ and $(3, f(3))$, and add text to label these points.

```
#define the function
f(x)=e^x

#generate the graph of the function
graph = plot(f(x),x,-0.5,3.5)

#generate the secant line
secant = line([(1,f(1)),(3,f(3))], color='black')

#add text to label these points
text1 = text("x=1", (1, f(1)+4), fontsize=20, color='red')
text2 = text("x=3", (3, f(3)+4), fontsize=20, color='red')

#display all in one
(graph+secant+text1+text2).show()
```

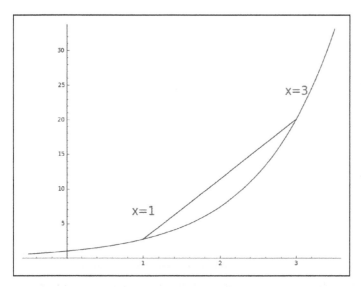

When we plot functions that have vertical asymptotes such as f(x)=tan(x) or $g(x) = \frac{1}{(x-1)*(x-2)(x-5)}$ then we obtain

```
plot(tan(x), -2*pi, 2*pi)
```

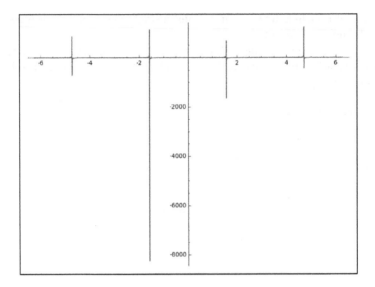

```
plot(1/((x-1)*(x-2)*(x-5)), -2*pi, 2*pi)
```

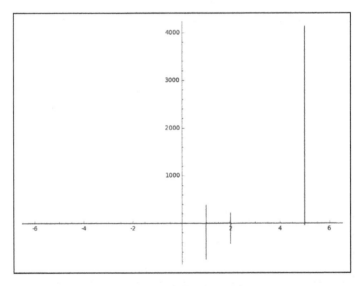

To address this issue, one can use the following options: `detect_poles=True` and `detect_poles='show'` together with the window size options to obtain

```
plot(tan(x), -2*pi, 2*pi, \
     detect_poles='show').show(ymin = -10, ymax = 10)
```

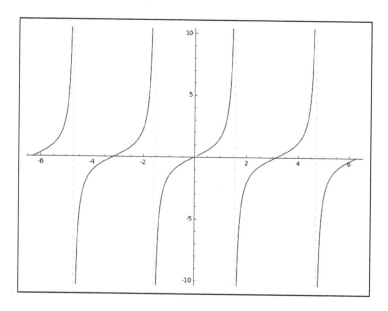

and

```
plot(1/((x-1)*(x-2)*(x-5)), -2*pi, 2*pi, \
    detect_poles='show').show(ymin = -10, ymax = 10)
```

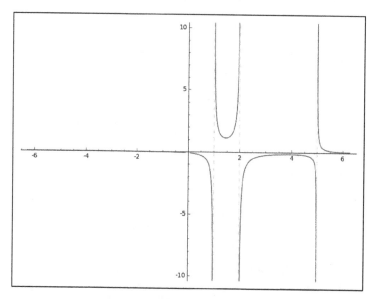

To add a grid to a graph, one can use the option `gridlines='minor'`:

```
plot(cos(x), -2*pi, 2*pi,  gridlines='minor').show()
```

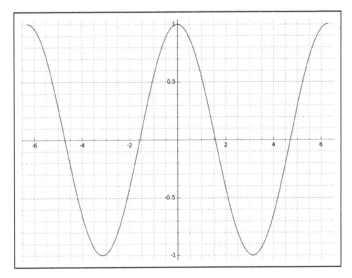

We next give an example that adds a set of lines and circles to a graph. We give more details about how to use a 'for' loop in the next chapter.

```
p = sum(line([(0,0),(cos(i),sin(i))])
              for i in [0,pi/20,..,2*pi])
q = sum(circle((0,0), i, color = 'red',
              thickness = 3)
              for i in [0,0.2,..,1])
(p+q).show()
```

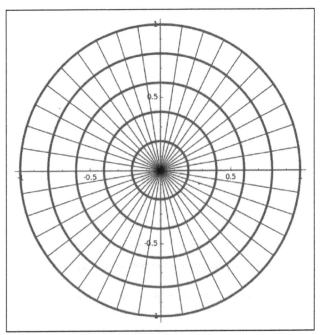

Sage Math can also plot parametric functions. For example,

```
parametric_plot( (cos(x), sin(x)*cos(x)), (x, 0, 6*pi),
                    color='blue',fill=True, fillcolor='red')
```

will give

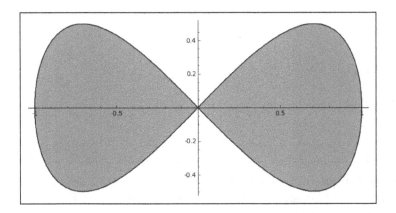

For more details on how to generate graphs of functions, points, lines, circles, and so on, the reader may want to read [16].

The last example in this section will be very useful in the Numerical Analysis examples that we'll see in the last chapter of this book. It will demonstrate how one can plot a given list of points. In here we generate 10 random points (each with coordinates between 1 and 6, inclusively), and then we plot them. Note the options `color` and `size` that are being used with `list_plot`. If you get duplicate points, then you will notice fewer points on the graph.

```
#we need the random library to have access to 'randint' function
import random

#random.randint(1,6) generates a random integer from[1,6]
#we create a list of points here
list_pts = [(random.randint(1,6),
            random.randint(1,6))
                for i in range(10)]

#we plot the previously created list
p = list_plot(list_pts ,color='red', size = 50)
p.show()
```

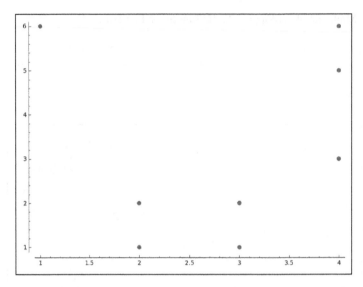

To label graphs, there are several options available. One can use the `title`, `legend_label`, or `axes_labels` options as explained next. For this, let's compare the following examples:

This first example has no labels:

```
f(x) = 1+x/2+x^2/2+x^3/6+x^4/24
plot(f(x), x, -1, 1)
```

and we obtain

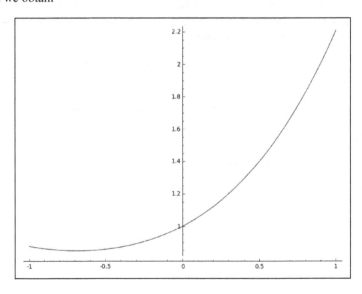

The next example uses the `title` option:

```
f(x) = 1+x/2+x^2/2+x^3/6+x^4/24
plot(f(x), x, -1, 1, title = f(x))
```

We obtain

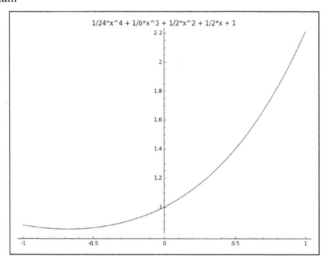

Alternatively, we get the same result if we convert the title name into a string:

```
f(x) = 1+x/2+x^2/2+x^3/6+x^4/24
plot(f(x), x, -1, 1, title = str(f(x)))
```

If we add "$" before and after the title (**just like in LaTeX for the Math mode!**)

```
f(x) = 1+x/2+x^2/2+x^3/6+x^4/24
plot(f(x), x, -1, 1, title = "$"+str(f(x))+"$")
```

we obtain a much nicer labeling:

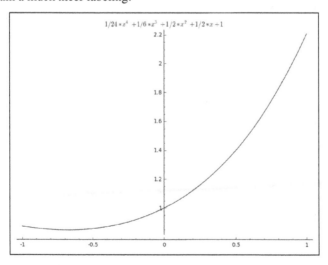

The `title` option accepts Math expressions that follow the LaTeX syntax with the observation that '`\`' needs to be doubled in order to avoid confusion with the escape characters used with strings such as '`\t`' (horizontal tab) and '`\n`' (newline):

```
f(x) = 1+x/2+x^2/2+x^3/6+x^4/24
plot(f(x), x, -1, 1,
     title = "this is nonsense but ... "+ \
     "$\\int_0^\\infty e^{-x^2} dx = \\frac{\\sqrt{\\pi}}{2}$")
```

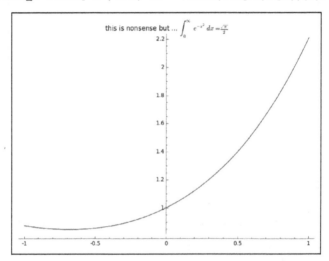

Another option we mention in here is the use of the `legend_label` option:

```
f(x) = 1+x/2+x^2/2+x^3/6+x^4/24
plot(f(x), x, -1,1, legend_label = str(f(x)))
```

We obtain

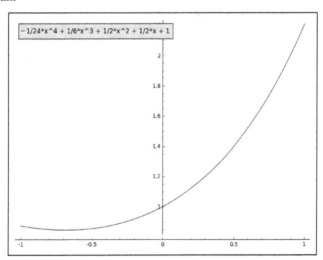

And the better alternative

```
f(x) = 1+x/2+x^2/2+x^3/6+x^4/24
plot(f(x), x, -1,1, legend_label = "$"+str(f(x))+"$").show()
```

to get

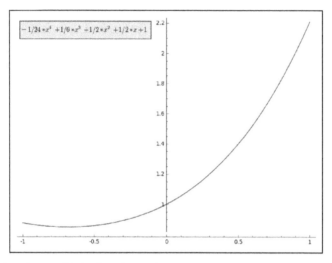

When drawing multiple functions, the `title` option is not very helpful, since only the last value of it will show up. For example,

```
f(x) = 1+x/2+x^2/2+x^3/6+x^4/24
g(x) = e^x
plot(f(x), x, -1,1, title = "$"+str(f(x))+"$") + \
plot(g(x), x, -1,1, title = "$"+str(g(x))+"$")
```

only displays the last value of `title`

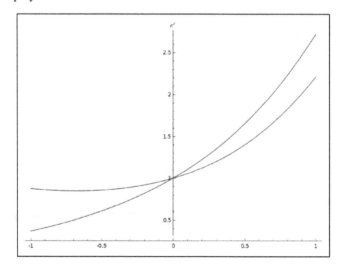

whereas the use of the `legend_label` option

```
f(x) = 1+x/2+x^2/2+x^3/6+x^4/24
g(x) = e^x
plot(f(x), x, -1,1, legend_label = "$"+str(f(x))+"$") + \
plot(g(x), x, -1,1, legend_label = "$"+str(g(x))+"$")
```

has better results

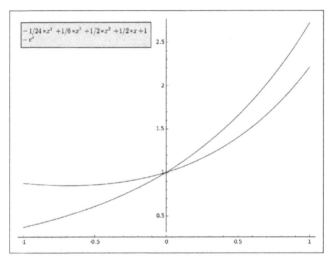

Moreover, different colors of the graphs will reflect in the legend:

```
f(x) = 1+x/2+x^2/2+x^3/6+x^4/24
g(x) = e^x
plot(f(x), x, -1,1, legend_label = "$"+str(f(x))+"$", color = 'red') + \
plot(g(x), x, -1,1, legend_label = "$"+str(g(x))+"$")
```

as one can note from the next image

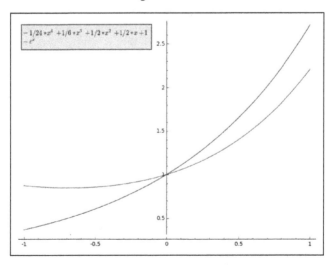

Not only different colors, but a different thickness or a different linestyle will reflect differently in the legend of the graph:

```
f(x) = 1+x/2+x^2/2+x^3/6+x^4/24
g(x) = e^x
plot(f(x), x, -1,1, legend_label = "$"+str(f(x))+"$",      \
                        color = 'red', linestyle='-.') + \
plot(g(x), x, -1,1, legend_label = "$"+str(g(x))+"$",      \
                        thickness = 4)
```

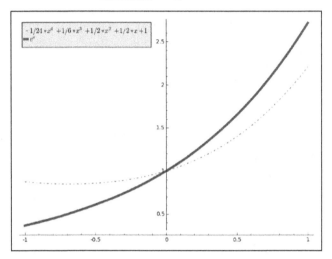

The last option we mention in here is `axes_labels`, which can be used in order to label each of the two axes. Here one needs to supply an array with two values, one for the label of each axis:

```
f(x) = 1+x/2+x^2/2+x^3/6+x^4/24
plot(f(x), x, -1,1, axes_labels =[ 'x',"$"+ str(f(x))+"$"])
```

We get

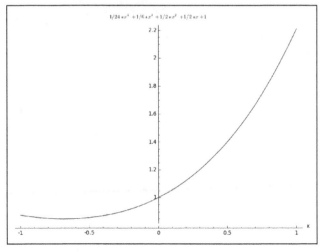

To plot a matrix, one can use the function `matrix_plot`. Using the option `cmap`, one can specify different coloring maps available (e.g., "Accent," "cool," "gnuplot," "ocean," "prism," "Purples," "Spectral," "terrain," and "winter"). One can also use the option `colorbar` (set to either `true` or `false`) and obtain a vertical color bar on the right side. Here are some examples. We are creating a 100×100 matrix and then plot it using different coloring maps.

```
#create a 500x500 matrix
A = matrix( [    [row*col for col in [1,2,..,500]]  \
             for row in [1,2,..,500]])

#plot a matrix
matrix_plot(A,colorbar=True, cmap='Accent',\
         aspect_ratio='automatic').show()
```

displays

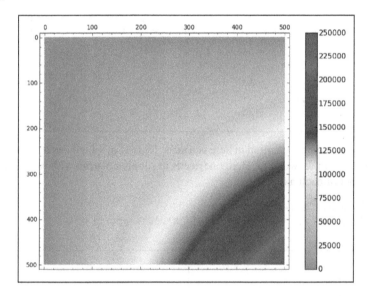

Changing the value of the option `cmap` to `'terrain'` one obtains

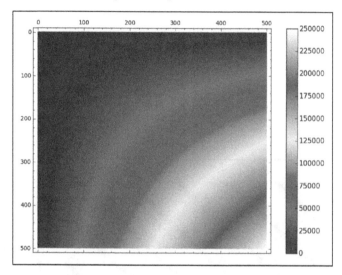

With some tweaks in this code, one can easily obtain some nice fractal-looking graphs:

```
#create a 500x500 matrix
A = matrix( [    [ (row*col)%100 for col in [1,2,..,500]]   \
                for row in [1,2,..,500]])

#plot a matrix
matrix_plot(A,colorbar=True, cmap='terrain',\
          aspect_ratio='automatic').show()
```

We obtain

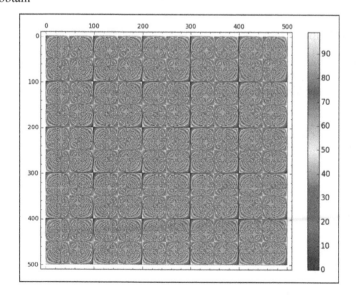

One last example:

```
#create a matrix
A = matrix( [    [ sin(row+col) for col in [0,pi/50,..,10*pi]]    \
                 for row in [0,pi/50,..,10*pi]])

#plot a matrix
matrix_plot(A,colorbar=True, cmap='gnuplot',\
            aspect_ratio='automatic').show()
```

which outputs

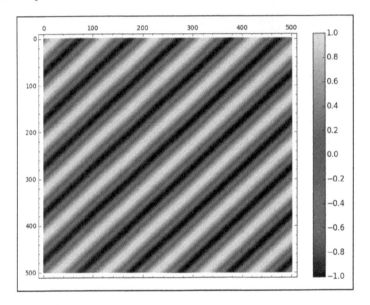

2.3.2   3D Graphs

Using the plot3d command, one can plot 3D graphs. Since we won't use this capability in this book, we will be rather brief in here.

Here is a Sage Math example of a 3D plot.

```
var("x,y")
f(x,y) = x + cos( x*y )
plot3d(f, (x,1,10), (y,1,10))
```

which generates

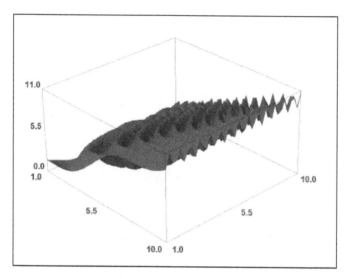

The great thing about 3D plots in Sage Math is that one can interactively rotate them. Doing so, we can get to see the plot from different angles. Here is an example (the same plot as the one above, viewed from a different angle).

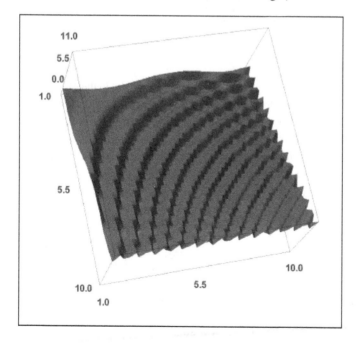

One can also zoom in/out the plot.

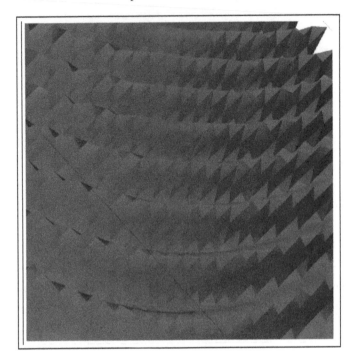

2.3.3 Exercises

(1) Graph the functions $f(x) = \sin(x)$ and $g(x) = \cos(x)$ in the same set of axes. Then find the smallest solution of the equation $\sin(x) = \cos(x)$ in the interval $[-\pi, \pi]$.

(2) Graph the functions $f(x) = \sin(x)$ and $g(x) = \cos(x)$ in the same set of axes. Then find the smallest solution of the equation $\sin(x) = \cos(x)$ in the interval $[0, 2\pi]$.

(3) Graph the functions $f(x) = x^2$ and $g(x) = e^x$ in the same set of axes. Then find the solution(s) of the equation $f(x) = g(x)$.

(4) Graph the functions $f(x) = x^3$ and $g(x) = e^{2x}$ in the same set of axes. Then find the solution(s) of this equation $f(x) = g(x)$.

(5) Graph the following parametric function:

$$\begin{cases} x = \sin(3t) \\ y = \cos(t + \sin(3t)) \end{cases}, t \in [0, 2\pi]$$

(6) Graph the following parametric function:

$$\begin{cases} x = \sin(t) \\ y = \cos(3t + \cos(3t)) \end{cases}, t \in [0, 2\pi]$$

(7) Graph the following parametric function:

$$\begin{cases} x = t * \sin(t) \\ y = t * \cos(t) \end{cases}, \quad \text{for } t \in [0, 2\pi]$$

(8) Graph the following parametric function:

$$\begin{cases} x = t * \sin(t) \\ y = t * \cos(t) \end{cases}, \quad \text{for } t \in [0, 4\pi]$$

(9) Create a list with 100 randomly generated points with both coordinates in the interval [20, 70] and plot them.

(10) Create a list with 200 randomly generated points with both coordinates in the interval [1, 6] and plot them.

(11) Create a list with 50 randomly generated points with coordinates in the interval $[1, 6] \times [10, 50]$ and plot them.

(12) Create a list with 70 randomly generated points with coordinates in the interval $[20, 40] \times [10, 90]$ and plot them.

(13) Draw and shade the region bounded above by $y = e^x$, bounded below by $y = x$, and bounded on the sides by $x = 0$ and $x = 1$.

(14) Draw and shade the region enclosed by the parabolas $y = x^2$, and $y = 2x - x^2$.

(15) Draw and shade the region enclosed by the curves $y = \dfrac{x}{\sqrt{x^2+1}}$, and $y = x^4 - x$.

(16) Draw and shade the region enclosed by the curves $y = \sin x$, $y = \cos x$, $x = 0$, and $x = \dfrac{\pi}{2}$.

(17) Draw and shade the region enclosed by the line $y = x - 1$, and the parabola $y^2 = 2x + 6$.

(18) Create a 3D plot for the function

$$f(x, y) = x^4 - y^4, \text{ in } [-10, 10] \times [-10, 10].$$

(19) Create a 3D plot for the function

$$f(x, y) = x^4 + y^4, \text{ in } [-10, 10] \times [-10, 10].$$

(20) Create a 3D plot for the function

$$f(x, y) = \cos(x) - y * x.$$

3

INTRODUCTION TO PROGRAMMING IN SAGE

In order to be able to write effective Sage Math code and implement your own Mathematical algorithms, you need to have a little bit of Sage programming background. The purpose of this chapter is to provide you this information. The next chapter will rely heavily on this one, so invest as much as needed now, and you'll have an easier time later. If you already master some other programming language, such as Python, Java, C/C++, or Ada, then you will most likely find this chapter rather easy.

If you have already known Python, then you may choose to skip this chapter entirely or merely glance at it. For a more comprehensive book about programming in general, and using Python in particular, the reader may want to search for the following great source [9]. Topics that are assumed to be known before reading this book (and one can find them in the above-mentioned reference) are binary numbers, memory, algorithms, and how computers work. This book is not a comprehensive programming textbook and in this chapter we introduce you briefly the main components used in implementing an algorithm: variables, decisional statements, loops, functions, and interacts. They should be sufficient in order to learn how to implement many algorithms, and in particular certain Sage Interacts (these will be covered more extensively in the next chapter).

We have seen in the previous two chapters that Sage Math has a rather rich library of functions and that we can use Sage as a Computer Algebra System. In this chapter, we learn how to implement our own algorithms and create our own Sage functions.

First thing one needs to learn in a specific programming language is how to use arithmetic expressions. We have already seen some Basic Arithmetic Expressions

An Introduction to SAGE Programming: With Applications to SAGE Interacts for Numerical Methods,
First Edition. Razvan A. Mezei
© 2016 John Wiley & Sons, Inc. Published 2016 by John Wiley & Sons, Inc.

earlier in this book (see p. (6) for a refresher). Next we will see a little more advanced programming statements.

3.1 VARIABLES

Here we learn about **variables**, which are named memory locations that can store values. The values can be numbers (a=3), strings (b = "Hickory"), functions (f=3*sin(x)), or more complex objects (p = plot(sin(x),x,0,2*pi)).

In the following example

```
steps = 10
```

we create a variable called steps and store the value 10 in it. To **store** a value, one needs to use the **assignment operator**, =. If you later need to change this value, let's say to 15, you can use the following:

```
steps = 15
```

and the value of the variable steps changes to 15. Variables are memory locations that can hold/store values, and you can change these values whenever needed (hence the name variable).

Note: It is important that you assign a value to a variable before trying to use it. The first time you assign a value to a variable is called **initialization**.

Note: Sage (just like Python, and many other programming languages) is case sensitive . This means that variables steps, Steps, sTePs, and STEPS will all refer to different variables. Try the following Sage code:

```
steps = 10
print STEPS
```

You will get the following error:

```
NameError: name 'STEPS' is not defined
```

When you choose a name for a variable you need to make sure it follows each of the following **variable naming rules**:

- It cannot contain spaces (e.g., number of spaces).
- The first character must be a letter or an underscore "_" (the following is illegal: 2steps).
- After the first character, you can only use letters, digits, and underscores.
- You cannot use keywords (words such as function, for, and while).

To output the value of a variable, we can use the print keyword.

For example, we can use the following:

```
steps = 10
print steps
```

to obtain

```
10
```

For a more user-friendly, we will make use of strings. A better, more informative way to output the value of steps would be

```
steps = 10
print 'steps =' , steps
```

and obtain

```
steps = 10
```

Note: A text that is contained between single quotes (' ') or double quotes (" ") is called a **string**. When you print **a string**, the output will be exactly what is in between the quotes (except for escape sequences – see p. (18)). So, for example,

```
print 'steps '
```

will output steps while

```
print steps
```

will output the **value of** steps, which in our example is 10.

Note: To output multiple expressions, we simply separate them by using commas. In our previous example, we wanted to output both, a string ("steps") and the value of a variable (the value of steps). Hence we used the Sage code: print 'steps =', steps.

Variables can store integers (whole numbers), float numbers (numbers involving the decimal point) strings (characters that are enclosed in quotation marks), or more advanced objects. As opposed to most other programming languages, in Python and in Sage Math one is allowed to reassign to a variable values of different types. For example, the following is legal in Sage Math (and Python):

```
steps = 10
print 'steps =' , steps
steps = 'none'
print 'steps =' , steps
```

and will give you the following output:

```
steps = 10
steps = none
```

Remember: A variable is a name that refers to some piece of data in the memory (i.e., volatile RAM memory), where some data is stored.

Note: Remember that the **assignment** operator is "=" whereas the **comparison** operator is "==." For example,

```
#assigns the value 10 to steps
steps = 10
#compares whether steps has the value 10
print steps == 10
```

outputs

```
True
```

Symbolic variables are used when we want to do symbolic computations. To declare them in Sage, we can use a syntax similar to

```
var('x,y,z')
```

or equivalently

```
var('x y z')
```

These statements declare x, y, and z as symbolic variables, and then they can be used in expressions such as

```
( (x+y)^2).full_simplify()
```

and obtain

```
x^2 + 2*x*y + y^2
```

To solve the quadratic equation

$$ax^2 + bx + c = 0,$$

one needs to declare a, b, c as variables first, and then use the solve method as shown below:

```
var('a,b,c')
solve(a*x^2+b*x+c==0,x)
```

We obtain

```
[x == -1/2*(b + sqrt(b^2 - 4*a*c))/a, x == -1/2*(b - sqrt(b^2 - 4*a*c))/a]
```

We will make extensive use of symbolic variables in symbolic functions (see p. 97). More details on these will be given throughout the book.

Note: The symbolic variable x is declared by default. Hence, one does not need to declare it. This simplifies the Sage code, especially when the only symbolic variable we need is x.

3.1.1 Exercises

(1) Which of the following variable names are illegal?
```
units_per_day
3DPlot
while
numberOfSteps
April2014
Appril2014
April2014.
April 2014
```
(2) Which of the following variable names are illegal?
```
number_of_steps
dayOfWeek
June 2016
2DPlot
function
month#3
Junne2016
June.2016
```
(3) Is the variable name `index` same as `InDeX`? Why or why not?
(4) What will the following Sage code output?
```
steps = 10
steps = 0
print "steps = ," steps
```
(5) Write a statement that will solve the equation $z^2 = 1$.
(6) Write a statement that will solve the equation $w^3 = 8$.
(7) Write a statement that will solve the equation $w^3 = a$.
(8) Write a statement that will solve the equation $z^2 = a^2$.

3.2 MORE ON OPERATORS

We have already seen some of the Sage arithmetic operators, such as `+`, `-`, `/`, `*`, `^`, `**`, `%`, `//`. One can make use of them when working with variables, working with constants (literals), or working with both. Below we introduce some more operators. They are very convenient and we make use of them in the following sections.

To add 4 to a variable:
One can use x = x + 4.
Equivalently, one can also use x += 4.

To subtract 4 from a variable:
One can use x = x - 4.
Equivalently, one can also use x -= 4.

To multiply a variable by 4:
One can use x = x * 4.
Equivalently, one can also use x *= 4.

To divide a variable by 4:
One can use x = x / 4.
Equivalently, one can also use x /= 4.
Other useful operators that we will not use in this book are //=, ^=, **=, and %=.

Note: It is important to note that there is no space in the operator +=. If you type + = instead of += a compiling error will be triggered. Typing

```
x = 1
x+ =10
```

will return the following error message:

```
SyntaxError: invalid syntax
```

Here is an example on how to use these operators:

```
x =  10
x +=  4
print 'Add 4 to x:', x

x =  10
x -=  4
print 'Subtract 4 from x:', x

x = 10
x *= 4
print 'Multiply x by 4:', x

x = 10
x /= 4
print 'Divide x by 4:', x

x = 10
x //= 4
print 'Divide x by 4:', x
```

```
x = 10
x^= 4
print 'Raise x to the 4th power:', x

x = 10
x **= 4
print 'Raise x to the 4th power:', x

x = 10
x %= 4
print 'Take x mod 4:', x
```

We obtained

```
Add 4 to x: 14
Subtract 4 from x: 6
Multiply x by 4: 40
Divide x by 4: 5/2
Divide x by 4: 2
Raise x to the 4th power: 10000
Raise x to the 4th power: 10000
Take x mod 4: 2
```

3.2.1 Exercises

(1) Write a statement that increases a variable x by 4.

(2) Write a statement that increases a variable x to 4 times its old value.

(3) Write a statement that doubles the value of a variable x.

(4) Write a statement that halves the value of a variable x.

(5) Write a statement that increases the value of a variable x by 10%.

(6) Write a statement that reduces the value of a variable x by 210%.

(7) Write a statement that computes the square root of a value x.

(8) Write a statement that computes the cubic root of a value x.

(9) Write a statement that computes the number of days in 2015 hours.

(10) Write a statement that computes the number of hours in 2015 minutes.

(11) Write a statement that computes the number of hours in 2,015,000 seconds.

(12) Write a statement that computes the number of days in 2,015,000 seconds.

3.3 MAKING DECISIONS

3.3.1 Boolean Expressions

An expression that evaluates to either True or False is called a **boolean expression**.

Relational operators such as > ("is greater than"), < ("is less than"), <= ("is less than or equal to"), >= ("is greater than or equal to"), == ("is equal to"), ! = ("is not equal to") they are used to obtain boolean expressions.

For example,

```
20>=27
```

evaluates to

```
False
```

The Sage code

```
x = 20
y = 27
x >= y
```

also evaluates to

```
False
```

To check whether two values are equal, one can use Sage code such as

```
x = 20
y = 27
x == y
```

and obtain

```
False
```

To check whether two values are different, one can use Sage code such as

```
x = 20
y = 27
x != y
```

and obtain

```
True
```

To build more complex boolean expressions, one can use the **logical operators**: and, or, and not.

- The and operator evaluates to true if both operands are true. It evaluates to false otherwise.

- The `or` operator evaluates to `true` if at least one of the operands is `true`. It evaluates to `false` otherwise.
- The `not` operator evaluates to `true` if the operand is `false`. It evaluates to `false` otherwise.

To check, for example, if $x \in [-20, 50)$ one can write a boolean expression such as

```
x>=20 and x<50
```

For example,

```
x = 25
x>=20 and x<50
```

evaluates to

```
True
```

To check whether or not $x \notin [-20, 50)$, one can either use Sage code similar to

```
x = 25
x<20 or x>=50
```

or to

```
x = 25
not ( x>=20 and x<50 )
```

and obtain, assuming x is 25:

```
False
```

Boolean values can also be obtained from function calls. For example,

```
2015.is_prime()
```

returns

```
False
```

Note how clear is to read code written in Sage, if you use good naming convention:

```
is_prime(2) and is_even(2)
```

which yields

```
True
```

How would one check whether the point $(4, 7)$ is on the line $y = 5x + 9$? One can use Sage code such as

```
7==5*4+9
```

or similarly, but maybe more clear

```
#we initialize the variables
x = 4
y = 7
#then we compare
y == 5*x+9
```

to obtain

```
False
```

which means the given point is not on the line $y = 5x + 9$.

3.3.2   If Statements

To write a piece of Sage code that is executed only when a condition (a boolean expression) evaluates to `true`, one needs to use an **if statement** (described below more fully). This way, some Sage statements will only be executed when certain conditions are satisfied (are evaluated to `true`).

For example, one can use the following Sage code that checks whether or not x is a prime number. Based on the value of this boolean expression, a specific message can be then printed out.

```
x = 2017
if(x.is_prime()):
    print "the number ", x, " is prime!"
else:
    print "the number ", x, " is NOT prime!"
```

outputs

```
the number 2017 is prime!
```

whereas a different value of x, let's say 2015

```
x = 2015
if(x.is_prime()):
    print "the number ", x, " is prime!"
else:
    print "the number ", x, " is NOT prime!"
```

outputs

```
the number 2015 is NOT prime!
```

We introduce Sage Interacts later in this chapter (see p. 103), but to pique your curiosity about them, and also to see a nice application to `if` statements, here is a nice example you may want to run in Sage:

```
@interact
def one_interact(
    n = slider(vmin=2, vmax=50, step_size=1, default=4, label="Select a num-
ber: ")):
    if(n.is_prime()):
        print n, "is prime!"
    else:
        print n, "is NOT prime!"
```

Selecting different values for *n* (using the slider provided), you will get different outputs:

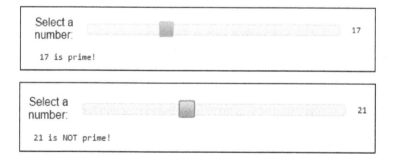

In the previous example, one should note the following syntax used for the **if–else statements**

```
if condition:
        true_block
else:
        false_block
```

The words `if` and `else` are **keywords**. That means they have a well-defined purpose (described here) that cannot be changed.

After the `if` keyword one needs to use a `condition` (any boolean expression, i.e., any expression that evaluates to either `true` or `false`), followed by a colon ("`:`").

If the `condition` evaluates to `true`, then the `true_block` is executed. The `true_block` is a set of one or more statements all being equally indented.

After the `true_block` comes the **else clause**. That contains the keyword `else`, followed by a colon, and then the `else_block`.

If the `condition` evaluates to `false`, then the `false_block` is executed. The `false_block` is a set of one or more equally indented statements.

A set of statements that are *equally indented* is called a **block of statements**.

Note: The **else** is optional. A simplified `if` statement would look like

```
if condition:
    true_block
```

In this case, if the `condition` evaluates to `true`, the `true_block` is executed. Otherwise, it is skipped.

Note: Make sure the `if` clause and `else` clause are <u>equally</u> indented.

In order to fully understand `if` statements, let's see more examples.

Write Sage code that given a number `x`, it outputs "is prime" if `x` is prime. Otherwise, it outputs a prime factorization of it.

```
x = 2015
if x.is_prime():
    print x, "is prime"
else:
    print x, "=", x.factor()
```

It outputs

```
2015 = 5 * 13 * 31
```

If one changes the value of `x` to 2017, then the output will be

```
2017 is prime
```

Write Sage code that computes the area of a rectangle, given the values of `width`, and `length`. If any of these values is negative, an error message should be displayed. Test your code for `width=3`, and `length=10`, and for `width=-3`, and `length=10`.

One can use the following Sage code:

```
width=3
length=10
if length<0 or width<0:
    print "the values should not be negative!"
else:
    area = length*width
    print "area =", area
```

and obtain

```
area = 30
```

changing the values of `width` and `length` to `-3` and `10`, respectively, one gets

```
the values should not be negative!
```

Write Sage code that checks whether or not a number is even. Based on the answer, it prints out a corresponding message.

One could write the following Sage code:

```
number = 7
if number%2 == 0:
    print "number", number, "is even"
else:
    print "number", number, "is odd"
```

and get

```
number 7 is odd
```

Changing number to 10, one would obtain

```
number 10 is even
```

How would you write a Sage code that given a number `grade`, it outputs the corresponding letter grade? Use the following table to decide what letter grade corresponds to a given number:

grade	letter grade
90$^+$	A
[80, 90)	B
[70, 80)	C
[60, 70)	D
[0, 60)	F
< 0	error

One could (not recommended!) write Sage code similar to

```
grade = 89.5
if grade>=90:
    print "A"
if grade>=80 and grade<90:
    print "B"
if grade>=70 and grade<80:
    print "C"
if grade>=60 and grade<70:
    print "D"
if grade>= 0 and grade<60:
    print "F"
if grade <0:
    print "error!"
```

The output would be

$\boxed{\text{B}}$

If the grade is changed to −3, then the output would be

$\boxed{\text{error!}}$

What would the output be if the grade is 107?

It is very important to note that the previous example *is not very efficient*. For example, in the grade=89.5 (such as above), then Sage Math does the following:

- First, it checks if grade≥ 90. Since this is not the case, it skips the following statement: print "A".
- Then, it checks to see if grade is in [80, 90). Since this IS the case, it executes the following statement: print "B".
- Then, it checks to see if grade is in [70, 80). Since this is NOT the case, it skips the following statement: print "C".
- Then, it checks to see if grade is in [60, 70). Since this is NOT the case, it skips the following statement: print "D".
- Then, it checks to see if grade is in [0, 60). Since this is NOT the case, it skips the following statement: print "F".
- Then, it checks to see if grade is negative. Since this is NOT the case, it skips the following statement: print "error!".

So why this is not efficient? Once we know that the grade is a B, Sage should skip all the other cases. It is not worth checking the remaining cases.

Many beginning programmers write their code as the one given above, when they first learn about if–else statements. But as they progress, they learn a more efficient way. Next we will see a better, more efficient way to solve the same problem. Namely, by using else statements. Here is the Sage code:

```
grade = 89.5
if grade>=90:
    print "A"
else:
    if grade>=80 and grade<90:
        print "B"
    else:
        if grade>=70 and grade<80:
            print "C"
        else:
            if grade>=60 and grade<70:
              print "D"
            else:
                if grade>= 0 and grade<60:
                    print "F"
                else:
                    print "error!"
```

By making use of the `else` clause, Sage will skip all the other test cases once it finds a match. In particular, for grade = 89.5, Sage will determine that the grade is a B and will skip all the subsequent tests.

A simplified way of writing the previous code is by using the `elif` keyword, as below:

```
grade = 89.5
if grade>=90:
    print "A"
elif grade>=80 and grade<90:
    print "B"
elif grade>=70 and grade<80:
    print "C"
elif grade>=60 and grade<70:
    print "D"
elif grade>= 0 and grade<60:
    print "F"
else:
    print "error!"
```

Note: All of the previous three versions of the code will have the same exact output (for the same input). But the last two will be more efficient (in terms of the number of operations performed). The very last one is perhaps the easiest to read.

Here comes another example. Write a Sage Math program that checks whether a given `year` is a leap year. A year is a leap year if it is divisible by 4 but not by 100 or if it is divisible by 400.

One may write the following

```
year = 2015

if (year
    print year , " is a leap year"
elif (year%400 == 0):
    print year , " is a leap year"
else:
    print year , " is NOT a leap year"
```

and obtain

```
2015 is NOT a leap year
```

If you change the `year` to 2016, then you get

```
2016 is a leap year
```

Next we will give a Sage application that randomly generates two numbers from [10, 20], and randomly picks an operation (addition, subtraction, multiplication, or division) and outputs a simple Math question. Note the efficient use of comments throughout this code.

```
#we need this library to have access to random() function
import random

#randomly generate a number in [10,20]
number1 = random.randint(10, 20)

#randomly generate another number in [10,20]
number2 = random.randint(10, 20)

#randomly pick a number from {1,2,3,4}
operation = random.randint(1,4)

# if 1 is selected then use addition
if operation == 1:
    print number1, "+", number2, "= ?"

# if 2 is selected then use subtraction
elif operation == 2:
    print number1, "-", number2, "= ?"

# if 3 is selected then use multiplication
elif operation == 3:
    print number1, "*", number2, "= ?"

# if 4 is selected then use division
else:
    print number1, "/", number2, "= ?"
```

Running this code multiple times will yield different, random, outputs similar to

```
10 * 11 = ?
```

```
20 - 11 = ?
```

```
15 / 20 = ?
```

The next example is a Sage program that can be used to randomly choose a value from a given list of values:

```
#we need this library to have access to random() function
import random

#put the list of values you want to choose from
list  = [12, 15, 16, 20, 22]

#randomly generate a position
pos = random.randint(0, len(list)-1)

#output the selected value from the list
print "the random choice is:", list[pos]
```

If you run this code multiple times, you'll get something similar to

```
the random choice is: 22
```

```
the random choice is: 12
```

```
the random choice is: 12
```

For more details on lists, see p. 88.

3.3.3 Exercises

(1) Write a boolean expression to check whether or not x is a prime number greater than 200.

(2) Write a boolean expression to check whether or not x is a prime number less than or equal to 2015.

(3) Write a boolean expression to check whether the point $(10, 20)$ satisfies the equation $y = 20x + 10$.

(4) Write a boolean expression to check whether the point $(15, 25)$ satisfies the equation $y = 20x^2 + 3x + 10$.

(5) Write Sage code that given a number x from $\{0, 1, 2, 3, 4, 5, 6\}$, it outputs the corresponding day of the week: Sunday for $x = 0$, Monday for $x = 1$, Tuesday for $x = 2$, Wednesday for $x = 3$, Thursday for $x = 4$, Friday for $x = 5$, and Saturday for $x = 6$.

(6) Suppose today is Tuesday. Given a positive integer no_of_days, output the name of the day of the week that is exactly no_of_days days after Tuesday. For example, if no_of_days = 3, then the output should be Friday. If no_of_days = 11, then the output should be Saturday.

(7) Write a Sage program that checks whether a given variable, x, is in the interval $[10, 50]$.

(8) Write a Sage program that checks whether a given variable, x, is in the interval $[10, 50)$.

(9) Write a Sage program that checks whether a given variable, x, is in the interval $(10, 50]$.

(10) Write a Sage program that checks whether a given variable, x, is in the interval $(10, 50)$.

(11) Write a Sage program that checks whether a given variable, x, is in $(-\infty, 10)$ or $(50, \infty)$.

(12) Write a Sage program that checks whether a given variable, x, is in $(-\infty, 10]$ or $(50, \infty)$.

(13) Write a Sage program that checks whether a given variable, x, is in $(-\infty, 10)$ or $[50, \infty)$.

(14) Write a Sage program that checks whether a given variable, x, is in $(-\infty, 10]$ or $[50, \infty)$.

(15) Write a Sage algorithm that decides whether a given variable `letter` "is a vowel," or "is a consonant."

(16) Write a Sage algorithm that decides whether a given number x "is a digit" (if the value of x is either 0, 1, 2, 3, 4, 5, 6, 7, 8, or 9), or "is not a digit."

(17) Write a Sage program that output the values of three given numbers, `a`, `b`, `c`, in increasing order.

(18) Write a Sage program that output the values of four given numbers, `a`, `b`, `c`, `d`, in increasing order.

(19) A company sells a product for $50. Quantity discounts are given according to the following table:

Quantity	Discount (%)
0–9	0
10–19	5
20–39	10
40–79	15
80+	20

Write a Sage program that, for a given value of the variable `quantity`, displays the total amount of purchase before the discount is applied, the amount of the discount, and the total amount after the discount is applied.

(20) Write a Sage program that uses the values of the variables `no_of_pennies`, `no_of_nickels`, `no_of_dimes`, and `no_of_quarters`. The program should compute the total value these coins added together. If the total value is equal to $1, the program should congratulate the user. Otherwise, the program should output a message indicating whether the amount entered was more than or less than $1.

(21) Given an amount of x total cents, write Sage code that outputs the same value in dollars and cents. For $x = 270$ your code should output "2 dollars, and 70 cents."

(22) Given an amount of x total seconds, write Sage code that outputs the same amount of time in hours and minutes. For $x = 10,000$ your code should output "2 hours, 46 minutes, and 40 seconds."

(23) Given a set of two points $(x_1, y_1), (x_2, y_2)$, write Sage code that outputs the slope of the line that passes through these points. Hint: Account for the possibility that the line may be vertical!

(24) Given a set of two points $(x_1, y_1), (x_2, y_2)$, write Sage code that outputs the equation of the line that passes through these points. Hint: Account for the possibility that the line may be vertical!

(25) Given values of the coefficients a, b, write Sage code that outputs the solutions of the equation $ax + b = 0$. Hint: a may be 0.

(26) Given values of the coefficients a, b, c, write Sage code that outputs the solutions of the equation $ax^2 + bx + c = 0$. Hint: a may be 0.

3.4 LOOPS

In this section, we learn how one can write Sage code that will cause a set of statements to execute repeatedly. For this, we study **Repetition Structures** (or **Loops**). We learn about `for` loops (some that are similar to Python, and other that are Sage specific), and then we'll also study `while` loops.

Note: You should invest as much time as needed to fully understand loops. They will be used extensively throughout the remaining of the book. In the author's teaching experience, many students seem to be very challenged when they first learn about *loops* (regardless of the programming language used). Once you master these loops in one programming language, it is quite easy to master them in most other programming languages too.

3.4.1 For Loops

When you need to repeat a set of statements a specific number of times (a *known number of times*), then `for` loops are usually the best choice.

The syntax for a **for** loop is

```
for variable in given_range:
    block_statements
```

where `variable` is a variable that will be assigned every value, one by one, from the `given_range`. The loop will execute the `block_statements` for each value assigned to the `variable`. Usually (but not always) the values of `variable` are used in this block of statements.

Let's start with a simple example. Although this program is not the most useful Sage code, it will help us begin our work with loops.

Write Sage code that outputs 10 times the message "Sage Math is cool!". One way to do this, would be

```
print "Sage Math is cool!"
print "Sage Math is cool!"
print "Sage Math is cool!"
print "Sage Math is cool!"
print "Sage Math is cool!"
print "Sage Math is cool!"
print "Sage Math is cool!"
print "Sage Math is cool!"
print "Sage Math is cool!"
print "Sage Math is cool!"
```

We obtain the expected output:

```
Sage Math is cool!
Sage Math is cool!
Sage Math is cool!
Sage Math is cool!
Sage Math is cool!
Sage Math is cool!
Sage Math is cool!
Sage Math is cool!
Sage Math is cool!
Sage Math is cool!
```

A shorter and more convenient way to get the same output is using a *loop*. For example, one could use the following Sage code:

```
for step in [1,2,3,4,5,6,7,8,9,10]:
    print "Sage Math is cool!"
```

Here step is called a temporary variable used inside the for loop. It is called a *dummy variable*, or *index*. For each value from the set $\{1, 2, 3, 4, 5, 6, 7, 8, 9, 10\}$, Sage assigns this value to step, and executes the statement that follows this line and is indented. As we will see below, one could use a set of statements, not only one.

So in our example, step takes the value 1, and then Sage outputs "Sage Math is cool!".

Then, step takes the value 2, and executes again the statement on the next line. Therefore, Sage outputs again "Sage Math is cool!".

Then, step takes the value 3, and executes again the statement on the next line. Therefore, Sage outputs again "Sage Math is cool!".

And so on, with values $4, 5, \cdots, 9, 10$.

The previous example could also be written as

```
for step in [1,2,..,10]:
    print "Sage Math is cool!"
```

Note: This code works in Sage but not in standard Python!

In the previous example, we did not use the value(s) of step. Here is an example that makes use of these values.

Write a Sage program that prints out the squares of the first nine positive integers:

```
for step in [1,2,..,9]:
    print step, "squared = ", step^2
```

We obtain the output:

```
1 squared = 1
2 squared = 4
3 squared = 9
4 squared = 16
5 squared = 25
6 squared = 36
7 squared = 49
8 squared = 64
9 squared = 81
```

Note: The names of the dummy variables in each of the previous examples was step. Instead of this name, one can choose any other name, as long as it is a valid variable name.

Here is another example of a loop. Write Sage code that prints out all the prime numbers between 50 and 70, inclusively.

Although we don't know how big the output is, the number of steps is known. We will have to check each integer in [50, 70], and if it is prime, we simply output it. Here is the Sage code for this:

```
for number in [50, 51,..,70]:
    if number.is_prime():
        print number
```

We obtain the output:

```
53
59
61
67
```

An alternative to the code given above, but with the same exact output, is

```
for number in range(50, 71)
    if number.is_prime():
        print number
```

The function range(50,71) returns a list with all the integers in the interval [50, 71). If you want to use a list of all the integers between 10, and 70, inclusively, then you could use either range(10,71), or [10,11,..,70]. If you have a strong Python programming background then you may prefer to use the standard function range().

Next we give another example of a for loop. **Twin primes** are prime numbers whose difference is exactly two. For example, (3,5), (5,7), and (11, 13). How would you write Sage code that prints out all the twin primes less than 2015?

Here is a possible solution. We "loop through" all the numbers between 2 and 2015-2 inclusively. Then, for each one of these numbers (let's refer to each one of them as x) we check if both x and $x + 2$ are prime. If they both are prime, then the pair $(x, x + 2)$ are twin primes and we display the pair. Here is the Sage code:

```
for x in [2,3,..,2015-2]:
    if(x.is_prime() and (x+2).is_prime()):
        print "(",    x,    ",",    x+2,    ")"
```

It outputs

```
( 3 , 5 )
( 5 , 7 )
( 11 , 13 )
( 17 , 19 )
( 29 , 31 )
( 41 , 43 )
( 59 , 61 )
( 71 , 73 )
( 101 , 103 )
.... we skipped some lines ...
( 1721 , 1723 )
( 1787 , 1789 )
( 1871 , 1873 )
( 1877 , 1879 )
( 1931 , 1933 )
( 1949 , 1951 )
( 1997 , 1999 )
```

If instead of printing these values we simply want to count the number of such pairs, then we change the previous Sage code into

```
count = 0
for x in [2,3,..,2015-2]:
    if(x.is_prime() and (x+2).is_prime()):
        count += 1
print "we found", count, " pairs of twin primes"
```

to get

```
we found 61 pairs of twin primes
```

As one can see, we needed to use some variable (we called it count) that kept track of the number of twin pair found at any given time. Each time we found a twin pair we increased count by one and then we moved on and checked the next values. At the end of the for loop, we displayed the value of count.

A combination of the previous two Sage algorithms would be the following:

```
count = 0
for x in [2,3,..,2015-2]:
    if(x.is_prime() and (x+2).is_prime()):
        count += 1
        print count, "\t(",    x,    ","  ,    x+2,    ")"
print "we found", count, " pairs of twin primes"
```

which prints each of the twin pair of primes along with an index.

Next we'll see how one can write a Sage code that computes a sum. For example, let's create an algorithm that will calculate

$$S = \sum_{i=10}^{2016} i^3.$$

In order to create the Sage algorithm, let's first see how one would solve this problem without using anything but paper and pencil. If you would have to compute this using paper and pencils what would you do? You could start with $S = 0$. Then add to it 10^3, then add 11^3, then 12^3, and so on until you finish adding 2016^3. Then, you simply output the value of S. The Sage code follows the same exact steps:

```
S = 0
for i in [10,11,..,2016]:
    S = S + i^3

print "total sum is S = ", S
```

We obtain that

```
total sum is S = 4133641992471
```

An equivalent alternative Sage code is

```
S = 0
for i in range(10, 2017):
    S = S + i^3

print "total sum is S = ", S
```

The variable S is called an **accumulator**, which is a variable used to compute a **running total**.

We next give a slightly more elaborate problem than the one just given above. Write Sage code that computes the following sum:

$$S = \sum_{i=10,i=even}^{2016} i^3.$$

One could use the following code:

```
#S holds the running total
S = 0
#loop through all even no. between 10 and 2016 incl.
for i in [10, 12, .., 2016]:
    S = S + i^3

print "total sum is S = ", S
```

or

```
#S holds the running total
S = 0
#third (optional) parameter of range is the step
for i in range(10, 2017, 2):
    S = S + i^3

print "total sum is S = ", S
```

to obtain

```
total sum is S = 2068870905568
```

Note: One could also use the Sage function sum to get the same results as the ones given above. To compute a sum of the form:

$$S = \sum_{k=a}^{b} f(k)$$

one could use Sage code similar to

```
sum([f(k) for k in [a..b]])
```

or

```
sum([f(k) for k in [a,a+1,..,b]])
```

Therefore, the previous example could **equivalently** be rewritten as

```
S =sum([k^3 for k in [10,12,.., 2016]])
print "total sum is S = ", S
```

Next we will write Sage code that computes the sum of all prime numbers between 100 and 2015, inclusively. We will again use a running total, let's call it sum. As before, we first need to initialize sum to 0. Then, using a loop, we "go" through each of the numbers $100, 101, 102, \cdots, 2015$. If the number is prime, then we simply add it to the running total, sum.

```
sum = 0
for number in [100, 101, .., 2015]:
    if number.is_prime():
        sum +=  number
print "sum = ", sum
```

and obtain

```
sum = 280004
```

The next Sage code will find and output the number of prime numbers less than 2015.

```
count = 0
for number in [2, 3,..,2015]:
    if number.is_prime():
        count +=  1
print "the number of primes less that 2015 is", count
```

We obtain

```
the number of primes less that 2015 is 305
```

We give now another example. Write Sage code that outputs a conversion table from Fahrenheit to Celsius using the formula:

$$C = \frac{5}{9}(F - 32).$$

The table should include the following values in Fahrenheit: $0, 5, 10, 15, \cdots, 100$. Here is the Sage code for this table:

```
for f in [0,5,..,100]:
    print f, " in F = ", 5/9.0*(f-32), " in C"
```

or, for a better output

```
for f in [0,5,..,100]:
    print f, "\tin F =", ( 5/9*(f-32)).n(20), "\tin C"
```

and we get

```
0     in F = -17.778  in C
5     in F = -15.000  in C
10    in F = -12.222  in C
15    in F = -9.4444  in C
20    in F = -6.6667  in C
25    in F = -3.8889  in C
30    in F = -1.1111  in C
```

```
35    in F =  1.6667    in C
40    in F =  4.4444    in C
45    in F =  7.2222    in C
50    in F = 10.000     in C
55    in F = 12.778     in C
60    in F = 15.556     in C
65    in F = 18.333     in C
70    in F = 21.111     in C
75    in F = 23.889     in C
80    in F = 26.667     in C
85    in F = 29.444     in C
90    in F = 32.222     in C
95    in F = 35.000     in C
100   in F = 37.778     in C
```

3.4.2 Strings

Strings (see also p. 17) are objects that can hold a text. To get the number of characters that a string holds, one can use the function: len(). Given a string variable, let's call it str, one can get the length of this variable using the code len(str). Strings have an array-type structure. That means that given a string variable, let's call it str, then you can access its first character using the index 0: str[0]. To access the second character in this string, one can use: str[1]. To access the last character in this string, one can use: str[len(str)-1].

So given the string str="Sage", then len(str) is equal to 4, and one can access the characters of this string using: str[0], str[1], str[2], and str[3].

Two strings can be concatenated using the "+" operator.

```
str1 = "Sage"
str2 = " is Cool!"
print str1+str2
```

The output is

```
Sage is Cool!
```

To concatenate/append a number to a given string, one needs to convert the number into string and then concatenate the two strings. For example, the following Sage code

```
word1 = "Sage "
word2 = 10
print word1+str(word2)
```

outputs

```
Sage 10
```

Next, we'll give some applications to strings that involve for loops.

Given a string variable called word, write a Sage Math code that will create an new string, call it new_word, containing every second character of the word, and then it will output new_word.

The solution to this problem is very similar to the previous examples where we computed a running total. The following Sage code

```
word = "SAGE is an awesome tool to do programming"
new_word = ""
for position in range(0, len(word), 2 ):
    new_word = new_word + word[position]
print new_word
```

output

```
SG sa wsm olt opormig
```

Given a string variable called word, write a Sage code that will create a new string, called reverse, that contains the characters of word in reverse order. Then, output the word reverse.

```
word = "SAGE is an awesome tool to do programming"
reverse = ""
for position in [len(word)-1,len(word)-2,..,0]:
    reverse = reverse + word[position]
print reverse
```

The code yields the following:

```
gnimmargorp od ot loot emosewa na si EGAS
```

A **palindrome** is a phrase that reads the same backward and forward. Use the previous code to write a Sage program that checks whether a given string is a palindrome.

In order to convert a number into a string, one can use the str function as follows:

```
print "Hickory is gr"+str(8)
```

to obtain

```
Hickory is gr8
```

In the previous example, one had to convert the number 8 to a string before it could be appended to the string "Hickory is gr." Therefore, we used the function str that converted 8 to a string and then it allowed us to append it to another string.

If you are trying to run the code without using the str function as shown below:

```
print "Hickory is gr"+8
```

then you will get an error message similar to

```
TypeError: unsupported operand parent(s) for '+':
'< type 'str'>' and 'Integer Ring'
```

Note: As mentioned above, to convert a numeric value into a string, one can use the `str` function. If you are given a string and you need to convert it into a number, then you can use the `Integer` function as shown next:

```
print Integer("444")+5
```

and get

```
449
```

If the given string does not represent an integer, then an error message will be displayed when you attempt to run the Sage code. For example,

```
print Integer("44.4")+5
```

produces the following error message

```
TypeError: unable to convert '44.4' to an integer
```

The alternative to `Integer` function in here is `float`:

```
print float ("44.4")+5
```

and obtain

```
49.4
```

3.4.3 While Loops

When you need to repeat a set of statements as long as a specific condition stays `true`(an *unknown number of times*), then `while` loops are usually a better choice than `for` loops.

The syntax for a **while** loop is

```
while condition:
    block_statements
```

where `condition` is a boolean expression. As long as `condition` stays `true`, the loop will execute the `block_statements`.

Sage first evaluates the `condition`. If it is `true`, then the `block_statements` is executed. Then it evaluates the `condition` again. If it is `true` it executes again the `block_statements`, and so on. Once the condition is evaluated to `false`, it skips the `block_statements` and executes the statements that follow the `while` loop.

Note: It is important that `condition` will eventually change its value to `false`. Usually some part of the `block_statements` will change the value of the `condition` to `false`. Otherwise, the loop will never end and will result in an *infinite loop*. This is a rather common error among beginning programmers.

We next give an example. Suppose one wishes to compute the sum of all digits of a given number. Then `while` loops should be used, as the number of digits is unknown.

Here is the trick that will be used:

- To compute the last digit of `number`, one can use: `number%10`.
- To get rid of the last digit of `number`, one should divide it by 10 and only get the integer part, hence one can use `number//10`.
- We continue the previous two steps until `number` becomes 0.

For example, if `number`= 2015. Then `number%10` is 5, and `number//10 = 201`. We would add 5 to a `sum`, and continue the work to 201. Sum would be 5, for now.

Then, for `number`= 201, one has that `number%10` is 1, and `number//10 = 20`. We would then add 1 to a `sum`, and continue the work to 20. So far, the sum would be 6.

Then, for `number`= 20, one has that `number%10` is 0, and `number//10 = 2`. We would then add 0 to a `sum`, and continue the work to 2. So far, the sum would be 6.

Then, for `number`= 2, one has that `number%10` is 2, and `number//10 = 0`. We would then add 2 to a `sum`, and continue the work to 0. So far, the sum would be 8.

Once `number` becomes 0, we stop and output the `sum`, 8.

To implement this in Sage, one could use

```
number = 2015
sum = 0
while number>0:
    sum = sum + number
    number = number // 10
print "sum of all digits is", sum
```

and obtain

```
sum of all digits is 8
```

Changing the previous number to 123121242344535667912000192 we would obtain

```
sum of all digits is 85
```

We give another example of `while` loops next. Suppose you need to find the smallest positive integer n that satisfies the expression

$$\frac{1}{2^n + 10n + 6} < 10^{-7}.$$

How would you solve this problem?

One way to solve this problem is by trying out, one by one, each positive integer (starting with 1) and stop as soon as n satisfies the required inequality. Since we don't know how many steps, we need to try until we find our solution, a `while` loop is appropriate in this example. Here is the Sage code:

```
#we start with n = 1
n = 1

#as long as n does not satisfy the required condition
while 1/(2^n+10*n+6)>=10^-7 :
    #increase n, try the next one
    n = n+1

#once we get here means we find our solution
print n
```

and the corresponding output is

```
24
```

Note: It should be noted that every Sage code involving a `for` loop could be rewritten so it uses a `while` loop. For this, the dummy variable must be initialized before the `while` loop, and then updated to advance to the next value in the block of statements.

Given a string variable called `word`, write a Sage code that will create an new string, `new_word`, containing every second character of the `word`, and then it will output `new_word`.

```
word = "SAGE is an awesome tool to do programming"
new_word = ""
position = 0
while position< len(word) :
    new_word = new_word+ word[position]
    position = position + 2
print new_word
```

output

```
SG sa wsm olt opormig
```

Write Sage code that computes the following sum:

$$S = \sum_{i=10,i=even}^{2016} i^3.$$

One could use the following code

```
S = 0
i = 10
while i<= 2016:
    S = S + i^3
    i = i + 2
print "total sum is S = ", S
```

to obtain

```
total sum is S = 2068870905568
```

Note: It is very important that the condition of the loop eventually becomes false so that the while loop eventually stops. For example, in the previous code if you remove the line $i = i + 2$, then i will never pass 2016, and hence the condition will never become false. In this case, you will get in an **infinite loop**. Try the code:

```
S = 0
i = 10
while i<= 2016:
    S = S + i^3
print "total sum is S = ", S
```

How would you write a Sage Math program that, given a number, prints out the smallest prime number that is greater than number? For example, the smallest prime number greater than 8 is 11, and the smallest prime greater than 13 is 17. How would you implement a Sage algorithm for this?

The idea is to start with the next integer greater than the given number and test if it is prime. If it is, we are done, we found our prime number. If not, we simply check the next positive integer. And we keep on going until we find the prime number we are looking for. Here is the Sage implementation of this algorithm:

```
number = 1000
current = number + 1
while not (current.is_prime()):
    current = current+1

print "the smallest prime greater than ",\
        number, " is: ", current
```

which gives us

```
the smallest prime greater than 1000 is: 1009
```

3.4.4   Nested Loops

As we will see in the next few examples, loops can also be nested. One can have a
loop inside a loop. This means that for each iteration of the outer loop, the inner loop
is run completely.

One example that uses nested loops would be the following. Create a Sage program
that displays the time (minutes followed by seconds) starting with 1:00 until 14:59.
Note that for each minute a complete loop through each of the seconds $0, 1, 2, \cdots, 59$
is needed, suggesting the need of a loop (through the seconds $0, 1, \cdots, 59$) for each
minute. To iterate through each minute, another loop is required. Here is a possible
solution:

```
for min in [1,2,..14]:
    for sec in [0, 1,..,59]:
        print "minute = ", min
        print "second = ", sec
        print
        sleep(1.0)   #system sleep for 1 second
```

We obtained

```
minute = 1
second = 0

minute = 1
second = 1

minute = 1
second = 2

minute = 1
second = 3

minute = 1
second = 4

minute = 1
second = 5

minute = 1
second = 6
...SKIPPED LINES ...
minute = 1
second = 58

minute = 1
second = 59

minute = 2
second = 0

minute = 2
second = 1
...SKIPPED LINES ...
minute = 14
second = 59
```

Another nested loops application is given next. Write Sage code that produces the following pattern:

```
*
**
***
****
*****
******
*******
********
```

In this pattern, one can see that the task is to produce a line with one star, then one with two stars, then one with three stars, and so on, until we produce a line with eight stars. To do so, one needs a loop that takes us through the values one through eight.

```
for no_of_stars in [1,2,..,8]:
     #print no_of_stars here.
```

How can one produce a given number of stars? Using a `for` loop similar to the one below, one can create a given `no_of_stars`:

```
line = ""
for step in range(no_of_stars):
     line = line + "*"
print line
```

Putting the previous two pieces of code together, one obtains the following solution:

```
for no_of_stars in [1,2,..,8]:
     #print no_of_stars here.
     line = ""
     for step in range(no_of_stars):
          line = line + "*"
     print line
```

Note the indentation! Indentation is used in Sage (as in Python) to indicate blocks of statements. Adjacent statements with the same indentation are part of the same block of statements.

Next we give another application of nested loops. Write Sage code that outputs the following pattern:

```
#
 #
  #
   #
    #
     #
```

This is a similar problem such as the one above.

First, we need a loop that takes us through each line. For this, we use code such as `for line in [1,2,..,6]:`.

Then, for each `line`, we need to put a `line` number of spaces, followed by the "#" character:

```
str = ""
for char in range(line):
    str = str + " "
str = str + "#"
print str
```

Putting together this code one obtains

```
for line in [1,2,..,6]:
    str = ""
    for char in range(line):
        str = str + " "
    str = str + "#"
    print str
```

3.4.5 Lists

We have already used lists with several occasions in the previous pages. In here, we give more details about what lists are and how to use them effectively.

We start with a first basic example:

```
[1,2,..,10]
```

which returns a list $\{1, 2, 3, 4, 5, 6, 7, 8, 9, 10\}$.

Another way to produce a list is

```
range(7)
```

in this case, the returned list is $\{0, 1, 2, 3, 4, 5, 6\}$.

Next we give some more elaborate examples of lists.

Lists can be given a variable name, and then using the square bracket operator one can get access to its elements. For example,

```
even = [0,2,..,10]
print 'the first element is: ', even[0]
print 'the second element is: ',even[1]
print 'third element is:', even[2]
print 'the last element is:', even[ len(even)-1 ]
```

would output

```
the first element is:  0
the second element is:  2
third element is:  4
the last element is:  10
```

Note that one can use the `len()` function to obtain the length of the list. Then, using the name of the list, and the indices $0, 1, \cdots$, len $- 1$, one can access individual elements of the list.

Note: It is very important to note that the first element in the list has index 0 (which is very common to many programming languages such as C/C++ and Java) and therefore the last element in the list has index len -1, not len! It is a common error among beginning programmers to be "off by one" and try to access the element at the index len; you should avoid this mistake.

From this perspective, one can look at strings as a list of characters. One can use the `len()` function on strings, as well as the bracket operator to access individual characters from the string. As an example, for the string `str = 'Lenoir-Rhyne University'`, `str[0]` returns `'L'`, `str[1]` returns `'e'`, and so on. Since `len(str)` returns 23 the last character in this string can be accessed as `str[23-1]` or `str[22]` and it returns `'y'`.

To create a list of pairs, one can use parentheses. For example, one can create the following set of points:

```
points = [(1,1),  (2,3),  (4,5)]
```

In this case, the `len()` function will return 3, and `points[0]` will return the first pair. To obtain the two coordinates of the first pair, one needs to use code such as `points[0][0]`, and `points[0][1]`. Here is a more elaborate example

```
points = [(1,2),(3,4),(5,6),(7,8)]
print points
print "length =", len(points)
print "the first point has coordinates:", points[0]
print "the first point has x = ", points[0][0], 'y=',points[0][1]
print "the second point has coordinates:", points[1]
```

This outputs

```
[(1, 2),  (3, 4),  (5, 6),  (7, 8)]
length = 4
the first point has coordinates: (1, 2)
the first point has x = 1 y= 2
the second point has coordinates: (3, 4)
```

A similar work can be done in the case of a three-dimensional list:

```
points = [(1,2,3), (4,5,6),(7,8,9),(10,11,12)]
print points
print "length =", len(points)
print "the first point has coordinates:", points[0]
print "the first point has x = ", points[0][0], \
                         'y = ',points[0][1], \
                      'and z = ',points[0][2]
print "the second point has coordinates:", points[1]
```

This outputs

```
[(1, 2, 3), (4, 5, 6), (7, 8, 9), (10, 11, 12)]
length = 4
the first point has coordinates: (1, 2, 3)
the first point has x = 1 y = 2 and z = 3
the second point has coordinates: (4, 5, 6)
```

Here is an example on how a Sage function can be used to plot a given list of points:

```
points = [(1,2),(3,4),(5,6),(7,8)]
list_plot(points, color= 'red')
```

and obtain

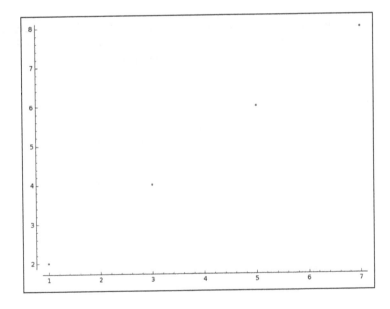

Just like with the strings, one can use the "+" operator in order to concatenate two lists. For example,

```
list1 = [1, 2, 5]
list2 = [3, 4]
list = list1+list2
print list
```

and obtain

```
[1, 2, 5, 3, 4]
```

One can also use a `for` **loop to create a list**. We'll use this quite frequently in the next chapter; therefore, you may want to invest some extra time (if needed) to understand this part. For example, the following creates a list of points and plots it.

```
points = [(i,i^2) for i in [1,2,..,6]]
list_plot(points, size = 60)
```

and obtain

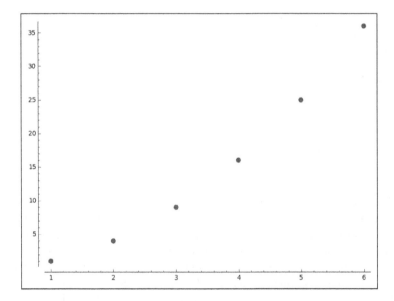

To create a list with the squares of the first 10 positive integers, one can use the following code:

```
points = [i^2 for i in [1,2,..,10]]
print points
```

and obtain

```
[1, 4, 9, 16, 25, 36, 49, 64, 81, 100]
```

One can use a random number generator to create a list. For example, the next Sage code will create a list of 20 randomly generated integers from [10, 70]

```
import random
mylist = [random.randint(10,70) for i in range(20)]
```

Given a list as the one given above, one could search for the largest value using the following algorithm:

```
#we initialize the variable max with
#the first element in the list
max = mylist[0]

#then we look for higher values in the list
for position in [1,2,..,len(mylist)-1]:
    #if we find something larger
    if max < mylist[position]:
        #we save it in max
        max = mylist[position]
#output the max
print 'the max value found is:', max

#print the list for verification
print mylist
```

We obtain

```
the max value found is: 62
[31, 43, 10, 17, 13, 15, 44, 62, 43, 60, 62, 48, 15, 51, 27, 59,
61, 28, 29, 39]
```

The above-given algorithm works as follows. Suppose you are given a (long) list of numbers. How do you find the largest value in it? You start with the first element in the list. Before you get to see the remaining elements that is the largest value and hence you store a copy of it in some temporary variable, let's call it max. Then, using a loop, you go through each element from the list and compare it against the value of max. If you find a value that is larger than max then save this value into max and continue your search through the list. Once you get to the end of the list, whatever value you have stored in max will be the largest value found in the list.

Next we give another example of using lists. The following program creates a list of 21 evenly spaced points from the interval [0,20] using the function $y = x\sin(x)$ to compute the corresponding y coordinates, and then it plots the newly created list.

```
mylist = [(x,x*sin(x)) for x in range(21)]
list_plot(mylist, size = 30)
```

It yields

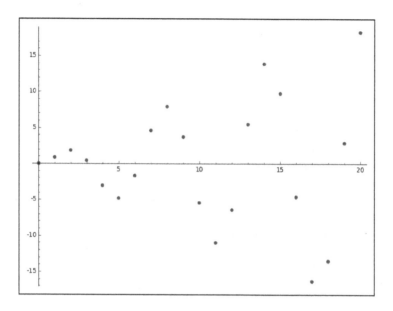

3.4.6 Exercises

(1) Write a Sage code that computes the following sum:

$$S_1 = \sum_{i=20}^{200} i^3.$$

(2) Write a Sage code that computes the following sum:

$$S_2 = \sum_{i=400}^{700} i^2.$$

(3) Write a Sage code that computes the following sum:

$$S_1 = 1^3 + 3^3 + 5^3 + \cdots + 2015^3.$$

(4) Write a Sage code that computes the following sum:

$$S_1 = 5^3 + 10^3 + 15^3 + \cdots + 2015^3.$$

(5) Write a Sage code that computes the sum of all prime numbers less than 2015.
(6) Write a Sage code that computes the sum of the squares of all prime numbers less than 2015.

(7) Write a Sage code that computes the product of all prime numbers less than 100.

(8) Write a Sage code that computes the product of all prime numbers between 1500 and 2015.

(9) Write a Sage code that computes the number of all prime numbers less than 100,000.

(10) Write a Sage code that computes the number of all prime numbers between 100,000 and 200,000.

(11) Use the Sage code given on p. 69 to create a test of 10 random algebra questions.

(12) Use the Sage code given on p. 69 to create a test of a random number of random algebra questions.

(13) Write a Sage program that counts the number of vowels in a given string variable.

(14) Write a Sage program that counts the number of consonants in a given string variable.

(15) A palindrome is a phrase that reads the same backward and forward. Use the code given on p. 81 to write a Sage program that checks whether a given string is a palindrome.

(16) Write a Sage program that changes every character of a given string into upper-case.

(17) In mathematics, $n!$ (to be read as n *factorial*) is computed using the formula:

$$n! = n \cdot (n-1) \cdot \cdots \cdot 3 \cdot 2 \cdot 1.$$

For $n = 0$, we define $0! = 1$. For $n < 0$, the factorial is undefined. Your task is to write a Sage code that computes $n!$ for any given value of n.

(18) Write a Sage code that computes the following:

$$P = \sum_{i=1}^{70} i^2.$$

(19) Given the values of the positive integers n, and k, $k \leq n$, write a Sage code that computes the following expression:

$$P = \frac{n!}{(n-k)!}$$

which represents the number of k-permutations from a given set of n elements.

(20) Given the values of the positive integers n, and k, $k \leq n$, write a Sage code that computes the following expression

$$C = \frac{n!}{k!(n-k)!}$$

which represents the number of k-combinations from a given set of n elements.

(21) Write a Sage program that checks whether or not a given number is square free. A number is **square free** if there is no perfect square, except 1, that divides the number.

(22) Write a Sage program that checks whether or not a given number is cube free. A number is **cube free** if there is no perfect cube, except 1, that divides the number.

(23) Use a while loop to check whether a given number is prime.

(24) Use a for loop to check whether a given number is prime.

(25) Write a Sage program that outputs a conversion table from Celsius to Fahrenheit using the formula

$$F = \frac{9}{5}C + 32$$

for the following values in Celsius: $\{-15, -10, -5, 0, 5, 10, \cdots, 40\}$.

(26) Write a Sage program that outputs a conversion table from centimeters to inches.

(27) Write a Sage program that outputs a conversion table from inches to centimeters.

(28) Write a Sage program that outputs a conversion table from liters to gallons.

(29) Write Sage code that outputs a 10×10 multiplication table.

(30) Write Sage code that outputs an 8×10 multiplication table.

(31) Create a list of 20 randomly generated points from $[10, 40]$.

(32) Create a list of 40 randomly generated points from $[-10, 20]$.

(33) Create a list of 20 randomly generated pairs of points from $[10, 40]$.

(34) Create a list of 40 randomly generated pairs of points from $[-10, 20]$.

(35) Create a list of 10 evenly spaced points from the interval $[0, 1]$.

(36) Create a list of 10 evenly spaced points from the interval $[-1, 1]$.

(37) Create a list of 20 evenly spaced points from the interval $[10, 100]$.

(38) Create a list of 20 evenly spaced points from the interval $[-50, 50]$.

(39) Create a list of 100 randomly generated points from the interval $[-100, 100]$ and then find the smallest value from the list.

(40) Create a list of 100 randomly generated points from the interval $[-100, 100]$ and then find the second smallest value from the list.

(41) Create a list of 100 randomly generated points from the interval $[-100, 100]$ and then find the sum of all the values from the list.

(42) Create a list of 20 randomly generated points from the interval $[-100, 100]$ and then find the product of all the values from the list.

(43) Create a list with 20 evenly distributed points of the graph of $y = \sin(x)$, on the interval $[-\pi, \pi]$ and plot them.

(44) Create a list with 20 evenly distributed points of the graph of $y = \cos(x)$, on the interval $[-\pi, \pi]$ and plot them.

(45) Write Sage code that produces the following pattern:

```
*
* *
* * *
* * * *
* * * * *
* * * * * *
* * * * * * *
* * * * * * * *
```

(46) Write a Sage code that produces the following pattern:

```
*
* * *
* * * * *
* * * * * * *
* * * * * * * *
* * * * * * * * * *
* * * * * * * * * * * *
* * * * * * * * * * * * * *
```

(47) Write a Sage code that outputs the following pattern:

```
x                 x
x               x
x             x
x         x
x     x
x x
x     x
x         x
x             x
x               x
x                 x
```

(48) Write a Sage code that computes the smallest value from a randomly generated list of 100 numbers.

(49) Write a Sage code that computes the second smallest value from a randomly generated list of 100 numbers.

(50) Write a Sage code that creates a randomly generated list of 100 numbers. Then it sorts the list and prints out the sorted list.

(51) Write a Sage code that creates a randomly generated list of 100 numbers. Then it sorts the list and it prints out the three smallest numbers.

3.5 FUNCTIONS

In Sage one can create functions either using a Python-style code or using a symbolic-style code. We will use both throughout this book.

Using functions not only you may make your code shorter but also you may make your code easier to read and easier to test, and it helps divide the work among several developers.

Here is an example using symbolic expressions:

```
f(x)=2*x+4
```

Then, one can use this definition in other expressions such as

```
print "f(5) =", f(5)
print "f(2*x) =", f(2*x)
print "f(cos(x^2))) =", f(cos(x^2))
print "f'(x)=", diff(f(x),x)
```

and obtain

```
f(5) = 14
f(2*x) = 4*x + 4
f(cos(x^2))) = 2*cos(x^2) + 4
f'(x)=2
```

One could also define a function using a Python-style:

```
def f(x):
    return 2*x+4

print "f(5) =", f(5)
print "f(2*x) =", f(2*x)
print "f(cos(x^2))) =", f(cos(x^2))
print "f'(x)=", diff(f(x),x)
```

and obtain the same exact output

```
f(5) = 14
f(2*x) = 4*x + 4
f(cos(x^2))) = 2*cos(x^2) + 4
f'(x)=2
```

A **function** is a block of statements that is given a name and can be executed upon request. It may or may not have parameters, and it may or may not return a value. Functions not only help us make the code shorter, but they also help us make the code more readable.

The syntax for defining a function is

```
def function_name( list_of_parameters):
    block_statements
```

Here function_name is a name that can be chosen using the same rules as the ones for variable names. In parentheses, it may use a set of parameters. These are values that are being sent by the part of Sage code that calls the function to execute. If there is no need for such values, then this list may be empty. Then the block_statements is the set of statements that we expect to be performed when we call the function to execute. If the caller of the function expects a result, then the value of this result can be sent back from the function using a return statement in the block_statements. This part is optional.

The block_statements is usually called the **body of the function**. All the statements from this block are indented.

Writing functions is an easy task, especially once you view some examples. Next we'll give plenty of examples and we will hopefully convince you that writing functions is not hard.

Next we give an example of writing functions and following it we provide more details about writing functions, local variables, scope, parameters, and so on.

Write a function that returns the square of a given number. Then use it to compute 5^2, 2015^2, and $(-0.5)^2$.

```
#defining the function square()
#as parameter we need to know the value to square
#let's call this value x
def squared(x):
        return x^2

#using the newly defined function
print squared(5), squared(2015), squared(-0.5)
```

The code outputs

```
25 4060225 0.250000000000000
```

Note: The parameters can be given any names that follow the rules of variable naming.

Note: Whenever a variable is created inside a function, this is called a **local variable**. This variable is not accessible outside of the function's body. If other functions use a local variable with the same name, these are unrelated, and they cannot "see" each other. The local variable can only be used inside the function it was created. That is the same as saying that the **scope** of the local variable is inside the function it was created.

Note: An **argument** is a value sent to a function when the function is called. For example, when we use the Sage code: squared(5), we are calling the function squared(), and we are sending the argument 5. The variable that receives this argument is called a **parameter** (or **parametric variable**). In our previous example, squared() function has only one parameter, x. We use this parameter inside the body of the function, in order to make use of the argument sent to the function.

Note: Any changes made to the parameter will not be saved to the original argument. This is because the local parameter will receive a copy of the argument's value.

Note: Some functions may have more than one parameter. In this case, whenever we call these functions, we should use the same order for the arguments as their corresponding parameters. There are other ways to send arguments to functions, but in this book we will only stick to this way (called *call by position*).

Note: As we mentioned above, a local variable is only accessible inside the function it was created. A variable that is created outside all functions in a program is called a **global variable**. It can be accessed by all functions, and hence shared across different parts/functions of the program. Because any part of the program can access a global variable, their use is discouraged as much as possible. They make debugging more challenging since any part of the code has access to it and can modify it.

Write a function print5() that outputs a given text (sent as an argument) five times.

```
def print5(text):
      print text
      print text
      print text
      print text
      print text
print5("Sage")
print5("MATH")
```

This will output

```
Sage
Sage
Sage
Sage
Sage
MATH
MATH
MATH
MATH
MATH
```

Can you improve the definition of print5() by using a for loop?

A Sage function add() that adds two values can be written as

```
def add(x, y):
     return x+y
```

Then, one can use it to add the values of two variables *a* and *b*, as follows:

```
a = 7
b = 12
print 'a + b =', add(a,b)
```

and obtain

```
a + b = 19
```

In this example, x, y are *parameters*, and a, b are *arguments*.

Next we will see an example where functions will be called inside other functions. Write a function `random_dice()` that will randomly draw a pair of dice, and output the result. Then create a function `game()` that will randomly draw seven pairs of dice and output these results.

Here is a possible solution:

```
import random
def random_dice():
    dice1 = random.randint(1,6)
    dice2 = random.randint(1,6)
    print "drawn: ", dice1, " and ", dice2

def game():
    for i in [1,2,..,7]:
        random_dice()

game()
```

We got the following:

```
drawn: 2 and 4
drawn: 1 and 6
drawn: 5 and 3
drawn: 1 and 3
drawn: 6 and 1
drawn: 4 and 1
drawn: 4 and 6
```

We can modify the previous example so that the output will also mention what is the draw number for each pair of die. For this we merely need to send a value from the `game` function to the `random_dice` and use it when we output the randomly generated values. Here is a possible solution:

```
import random
def random_dice(num):
    dice1 = random.randint(1,6)
    dice2 = random.randint(1,6)
    print "drawn #", num,": ", dice1, " and ", dice2

def game():
    for i in [1,2,..,7]:
        random_dice(i)

game()
```

We got the following:

```
drawn #1:  2  and  4
drawn #2:  1  and  6
drawn #3:  5  and  3
drawn #4:  1  and  3
drawn #5:  6  and  1
drawn #6:  4  and  1
drawn #7:  4  and  6
```

3.5.1 Using Library Functions: Random, SciPy, NumPy

To have access to pre-installed modules (or libraries), one needs to `import` them (unless they are imported by default by Sage –certain libraries are imported by default). This allows you to have access to functions, classes, and constants defined in these libraries.

A particular module containing several functions useful when dealing with (pseudo-)random number generators is `random`. In the example given right before this subsection, we used the following statement:

```
import random
```

This allowed us to use the function `random.randint` defined in the `random` library.

Other important modules that we will make use of in this book are `scipy` and `numpy`. SciPy is a Python-based set of open-source packages for Mathematics, Science, and Engineering. Among others, it includes the following core components: NumPy and SciPy library, Matplotlib, Sympy, and so on. We can use these libraries in Sage as well. Throughout this book, we will see several examples that will import these libraries and make use of their scientific computing functions:

```
import scipy
import numpy
```

For more details on these libraries than those provided by this book, you may want to access the official websites: [10], [14] See also [6].

3.5.2 Exercises

(1) Define a function `is_even()` that returns `true` if a given value is even, `false` otherwise.

(2) Define a function `is_palindrome()` that returns `true` if a given word is palindrome, `false` otherwise.

(3) Define a function `is_perfect()` that returns `true` if a given number is a perfect square, `false` otherwise.

(4) Define a function `is_prime()` that returns `true` if a given number is a prime number, `false` otherwise.

(5) Define a function `print2015()` that prints a given text 2015 times.

(6) Define a function `upper()` that prints a given text using only uppercase characters.

(7) Define a function `lower()` that prints a given text using only lower characters.

(8) Write a function `FtoC()` that returns the value in Celsius, of a given value in Fahrenheit. Use the formula

$$C = \frac{5}{9}(F - 32).$$

(9) Write a function `CtoF()` that returns the value in Fahrenheit, of a given value in Celsius. Use the formula

$$F = \frac{9}{5}C + 32.$$

(10) Write a function `letter_grade()` that returns the letter grade of a given score. For example, `letter_grade(98)` should return A, and `letter_grade(59)` should return F.

(11) Create a Sage function `random_question()` that will randomly pick two numbers between 0 and 20 and then will randomly pick a basic math operation (addition, subtraction, multiplication, or division). Then, the function will output the question similar to " 6 + 7 = ?." Create another function called `test()` that will print 10 random questions. Then call this function to generate two tests.

(12) Create your own function `solve_quadratic(a,b,c)` that can solve an equation of the form $ax^2 + bx + c = 0$. Then use it to find the solutions of the following equations: $x^2 - 6x + 5 = 0, x^2 + 1 = 0, x + 3 = 0$.

(13) Create your own function `solve_cubic(a,b,c,d)` that can solve an equation of the form $ax^3 + bx^2 + cx + d = 0$. Then use it to find the solutions of the following equations: $x^3 - 1 = 0, x^3 - 6x + 5 = 0, x^2 + 1 = 0, x + 3 = 0$.

(14) Create a function `factorial`, that computes the factorial of a given positive integer n using the formula

$$n! = n \cdot (n - 1) \cdots \cdots 3 \cdot 2 \cdot 1.$$

(15) Write a function that returns the conversion of a value from centimeters into inches.

(16) Write a function that returns the conversion of a value from inches into centimeters.

(17) Write a function that returns the conversion of a value from gallons into liters.

(18) Create a function `permutations()` that computes the following expression:

$$P = \frac{n!}{(n - k)!}$$

which represents the number of k-permutations from a given set of n elements ($n \geq k \geq 0$).

(19) Create a function `combinations ()` that computes the following expression:

$$C = \frac{n!}{k!(n-k)!}$$

which represents the number of k-combinations from a given set of n elements $(n \geq k \geq 0)$.

(20) Create a function `sort ()` that sorts a given list from the smallest to highest (ascending sorting).

(21) Create a function `sort ()` that sorts a given list from the highest to smallest (descending sorting).

(22) Create a function `adds ()` that adds the values of all elements from a given list of numbers.

(23) Create a function `countEven ()` that counts the number of values from a given list that are even numbers.

(24) Create a function `countPrimes ()` that counts the number of values from a given list that are prime numbers.

3.6 INTERACTS

Interacts are a great way to provide an interactive application to the user. In this section, we introduce some interactive examples using **input boxes, sliders,** and **selectors.** One can find many more examples of interacts in [16], in particular on `http://wiki.sagemath.org/interact/`. The next entire chapter is dedicated to Sage Interacts for Numerical Methods.

First, we'll introduce some basic examples that introduce each of **input boxes, sliders,** and **selectors.** Then we'll demonstrate some more elaborate examples. Then we present some other types of Sage Interacts: **check box,** and **range slider.**

Note: Note from all the examples given here that the Sage Interacts must start with `@interact`.

We'll start with the **input box.** An *input box* is used to get the user's input. For example,

```
@interact
def myFirstInteract(
    n = input_box(default = "Anonymous",
                  label = "Type your name here:",
                  type = str)):
        print("Welcome " + n + "!")
```

will output

If you change the text in the input box to "George" and hit enter, then you obtain

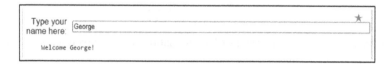

In the above example, n is the name of the parameter that is used to read the user's input, interacting with the input box. We could have used any valid variable name, in place of n. As one can see from this example, several options are available with an input box: a `default` value, a `label` for the input box, and the `type` of input to be read. The above-given interact was named `myFirstInteract`.

Another type of Sage Interact is a **slider**; it allows the user to pick a value from a predefined range. The following example

```
@interact
def mySecondInteract(
    n = slider(vmin=0, vmax=80, step_size=1,
               default=4, label="Select a number: ")):
        print "You selected: " ,   n
```

will output something similar to

Again, n is the given name to the parameter that is used to read in the user's input when it interacts with the slider. We could have used any valid variable name, in place of n. As one can see from this example, several options are available with a *slider*: a minimum value, a maximum value, the thicks used (or `step size`), the `default` value selected, and a `label` for the slider.

We give next another Sage Interact example that uses a *slider*. In here, we use the `Fibonacci()` function included in the Sage library to compute the nth Fibonacci number. Using a *slider*, we allow the user to select a value for n.

```
@interact
def FibonacciInteract(
    n = slider(vmin=0, vmax=80, step_size=1,
               default=4, label="Select a number: ")):
        print  str(n)+"th fibonacci number=" , fibonacci(n)
```

We obtain

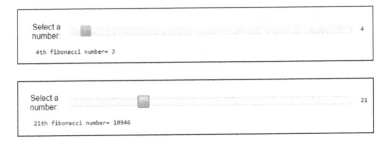

and

Select a
number: 67

67th fibonacci number= 44945570212853

The third type of a Sage Interact is a **selector**. It allows the user to select from a small (and well-defined) set of options that may or may not be numbers. Here is an example of a Sage Interact that uses a *selector*:

```
@interact
def myInteract3(
    n = selector(values = ["Triangle", "Square", "Pentagon", "Other"],
                 label = "Select a polygon: ",default = "Square" )):
    print "You selected: " ,   n
```

that will output something similar to

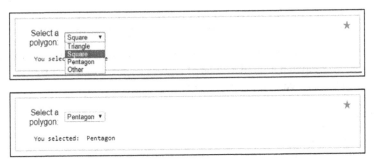

As with the previous examples, *n* is the name of the parameter that is used to read the user's input, interacting with the selector. There are several options available with a selector: a list of `values`, a `default` selected value, and a `label` for the selector.

More complex Sage Interacts will typically use a combination of sliders, input boxes, and selectors, not only one at a time. The next examples will be a display of some more elaborate and more useful interacts.

Next, we'll create a Sage Interact that computes the Taylor polynomial, around $c = 0$, of a given function. We'll use an input box to input a function, and a slider to pick the order of the Taylor polynomial.

```
@interact
def myInteract4(
    f = input_box(default=e^x ),
    n = slider(vmin=0, vmax=10, step_size=1,
               default=3, label="Select the order n: ")):
    print  f, " = "  , f.taylor(x, 0, n)
```

This outputs

If the user changes the function to $\sin(x^2)$ and selects order 7, then the output is

The following example will extend the previous Sage Interact so that the Taylor polynomial is computed around a value of c given by the user. For this, we'll use an *input box* so the user can input a value of c.

```
@interact
def myInteract5(
    f = input_box(default=e^x ),
    n = slider(vmin=0, vmax=10, step_size=1,
               default=3, label="Select the order n: "),
    x0 = input_box(default=0 )):
    print  f, " = "  , f.taylor(x, x0, n)
```

We obtain

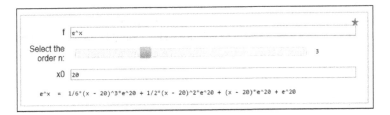

and

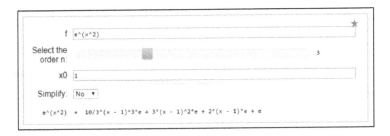

Note: The *input boxes* and the *slider* appear in the same order in which they were declared.

To the previous example, one can add a *selector* and allow the user to get a simplified Taylor polynomial, or a nonsimplified one.

```
@interact
def myInteract6(
    f = input_box(default=e^x ),
    n = slider(vmin=0, vmax=10, step_size=1,
               default=3, label="Select the order n: "),
    x0 = input_box(default=0 ),
    simplified = selector(values = ["Yes", "No"],
                 label = "Simplify: ",default = "No" )):
    if(simplified == "Yes"):
        print  f, " = "  , f.taylor(x, x0, n).full_simplify()
    else:
        print  f, " = "  , f.taylor(x, x0, n)
```

We obtain

and

In the previous code, we used the value of the selector to decide whether or not to apply `full_simplify()` to the obtained Taylor polynomial.

An alternative to a *selector* is a **check box**. This would be more appropriate when the user can only select between two values. To create a check box, one can use the following syntax:

```
boolean_variable = (label, default_value)
```

where `default_value` can be either `true` or `false`. One can rewrite the previous Sage Interact similar to

```
@interact
def myInteract7(
    f = input_box(default=e^x ),
    n = slider(vmin=0, vmax=10, step_size=1,
            default=3, label="Select the order n: "),
    x0 = input_box(default=0 ),
    simplified = ("Simplify:", True)):
    if(simplified == True):
        print  f, " = "  , f.taylor(x, x0, n).full_simplify()
    else:
        print  f, " = "  , f.taylor(x, x0, n)
```

to obtain

and

Here is another application of *check boxes*. This Sage Interact will allow the user to select which functions to be plotted. For each of the functions considered, the user can choose whether or not to display it.

```
@interact
def sineWaveInteract(
    x1 = ("1*2*pi", true),
    x3 = ("3*2*pi", true),
    x5 = ("5*2*pi", true),
    x7 = ("7*2*pi", true)):
    var('x')
    f = 0*x

    if(x1 == true):
        f  += sin(1*2*pi*x)

    if(x3 == true):
        f += 1/3* sin(3*2*pi*x)

    if(x5 == true):
        f += 1/5* sin(5*2*pi*x)

    if(x7 == true):
        f += 1/7* sin(7*2*pi*x)

    f(x) =f
    plot(f(x),0, pi).show()
```

We obtained

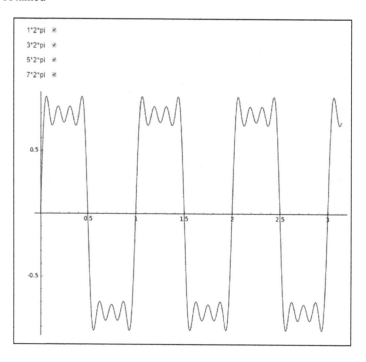

and

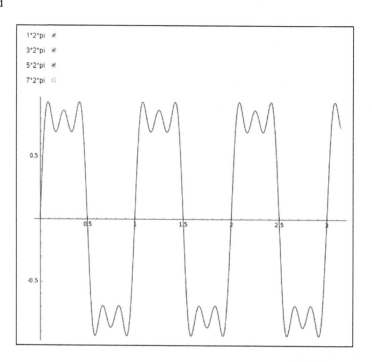

The previous example could be rewritten (not equivalently though!) to make use of a *slider*:

```
@interact
def sineWaveInteract(
    n = slider(vmin=1,vmax=11, step_size = 2)):
    var('x')
    f = sin(2*pi*x)

    for i in [3,5,..,n]:
        f  += 1/i * sin( i*2*pi*x)

    f(x)=f
    plot(f(x),0, pi).show()
```

and get

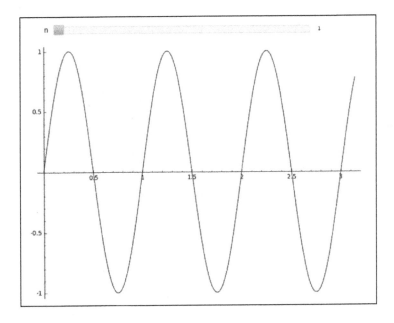

then, if the user chooses $n = 3$:

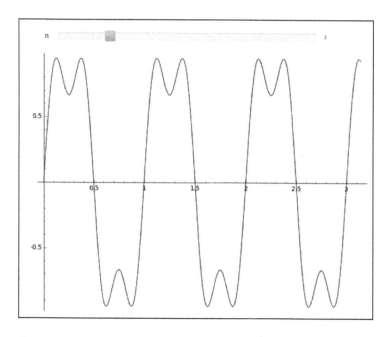

and, for $n = 11$:

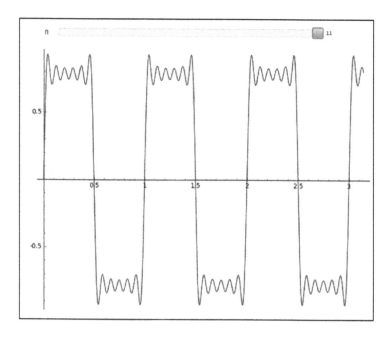

Sage Interacts also offer you the possibility to use a **range slider** in order to allow the user to specify an interval. The *range slider* has several options available, including the smallest , the largest, and the step of the interval, the default interval, and the option of including a label.

For example, the following Sage Interact outputs the Taylor polynomial of a selected function for all the integers from a selected range:

```
@interact
def myInteract10(
    f = input_box(default=e^x ),
    interval=range_slider(0, 10, 1, default=(0, 4),
                label="Select the range for n:"),
    x0 = input_box(default=0 ),
    simplified = ("Simplify:", True)):
    for n in range(interval[0], interval[1]):
        if(simplified == True):
            print  f, " = "  , f.taylor(x, x0, n).full_simplify()
        else:
            print  f, " = "  , f.taylor(x, x0, n)
```

We obtain

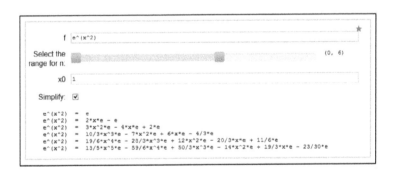

A simpler application for *range sliders* is given next. The following Sage Interact will allow the user to select a domain for plotting the function $f(x) = \cos(x)$.

```
@interact
def myInteract11(
    interval=range_slider(-6*pi, 6*pi, pi/4,
                    default=(0, pi),
                    label="Select the range for f(x):")):
    f(x) = cos(x)
    p = plot(f(x),x, interval)
    p.show()
```

We obtain

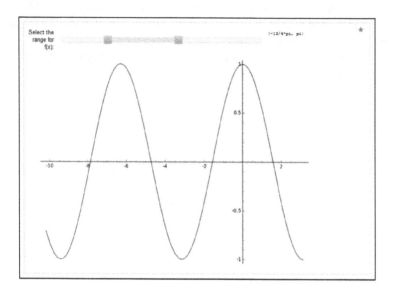

and

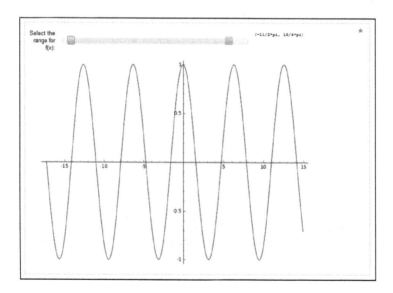

Next, we'll develop a Sage Interact that will accept an input number, in base 10, and it will output the base 2 representation of the chose input:

```
@interact
def Base10To2(n = input_box(2015)):
    #we will convert n from base 10 into base 2
    tmp = n
    base2 = ""
    while(tmp!=0):
        base2 = str( tmp
        tmp //= 2
    print n, "in base 10 = ", base2, "in base 2"
```

and we obtain

```
        n  2015

        2015 in base 10 =  11111011111 in base 2
```

```
        n  3

        3 in base 10 =  11 in base 2
```

and

```
    n  123213234342342

    123213234342342 in base 10 =  111000000001111101000100010000100010011100011 in base 2
```

We challenge you to change this Sage Interact so it uses a *slider* instead of an *input box*.

By default, the Sage Interact will automatically update as soon as you change a value in an *input_box*, or a *slider*, and so on. If you want to turn off this feature, you can use the auto_update option set to False. Here comes an example:

```
@interact
def UpdateInteract(
    f=input_box(e^x),
    a = input_box(0),
    b = input_box(2),
    auto_update=False):
    f(x)=f
    p = plot(f(x), x, a,b)
    p.show()
```

This Sage Interact will have the following output:

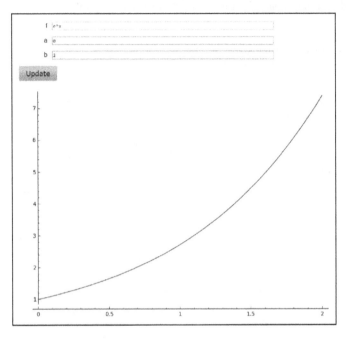

If you change any value, a red rectangle will notify you that changes occurred in certain places and a click on the "Update" button is needed in order to refresh the output

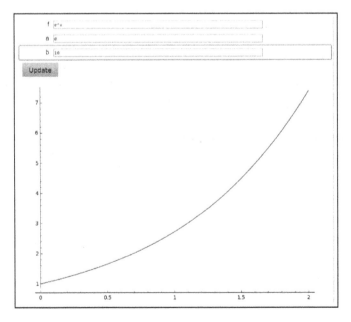

When the Sage Interacts involve lengthy computations that are time consuming, the Update option is very handy. It will allow the user to input or select all the needed parameters before updating the output.

In order to create even more complex applications, one can also use **nested interacts**. These are a little more complex, but the outcome can be very rewarding. For example, the following Sage Interact will first read a set of data points, called `points`. Then, based on the number of points entered, it will create a *slider* that will allow the user to select and output one element at a time. These elements can be numbers, ordered pairs, triples, and so on.

```
@interact
def OuterInteract(
    points    = input_box([[(1,1), (2,2), (3,1.5), (4,4)]] ) ):

  n = len(points)

  @interact
  def NestedInteract( position = slider(vmin=0,
                                  vmax=n-1,
                                  step_size = 1,
                                  default = 1) ):
      print "at position", position,        \
            "we have the point:", points[position]
```

We obtain

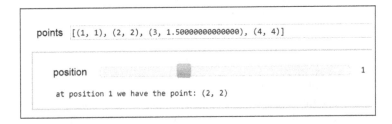

```
points  [(1, 1), (2, 2), (3, 1.50000000000000), (4, 4)]

position                                              1
at position 1 we have the point: (2, 2)
```

and, changing the points to 1, 2, 4, 7, 8, 9, 100, we can see that the maximum position that we can select is 6:

```
points  1, 2, 4, 7, 8, 9, 100

position                                              6
at position 6 we have the point: 100
```

The following input works too: `[(x,x^2,x^3) for x in [5,` `8,..,2015]]`, and we get

```
points  [(x,x^2,x^3) for x in [5, 8,..,2015]]

position                                                          635

at position 635 we have the point: (1910, 3648100, 6967871000)
```

3.6.1   Exercises

(1) Create a Sage Interact that allows the user to enter a value and outputs whether the number inserted is prime.

(2) Create a Sage Interact that allows the user to enter a value and outputs a prime factorization of the number.

(3) Create a Sage Interact that asks the user to input two positive integers n and k. If $0 \leq k \leq n$, output the value of n choose k using the formula:

$$C = \frac{n!}{k!(n-k)!}.$$

Otherwise display an error message.

(4) Create a Sage Interact that asks the user to input two positive integers n and k. If $0 \leq k \leq n$, output the value of

$$P = \frac{n!}{(n-k)!}.$$

Otherwise display an error message.

(5) Create a Sage Interact that allows the user to input a function, and select a positive integer n less than 10. Then, it plots the function and the Taylor polynomial of degree n.

(6) Extend the previous Sage Interact so that the user can choose a color for plotting the graph of the function and a color for plotting the graph of the Taylor polynomial.

(7) Create a Sage Interact that allows the user to input a text and outputs whether the text introduced is a palindrome.

(8) Create a Sage Interact that allows the user to input a text and it displays reversed.

(9) Create a Sage Interact that allows the select (using a slider) different values of n, and then it outputs the sum of:

$$\sum_{k=1}^{n} \frac{1}{k}.$$

(10) Create a Sage Interact that allows the select (using a slider) different values of n, and then it outputs the sum of:

$$\sum_{k=1}^{n} \frac{1}{k^2}.$$

(11) Create a Sage Interact that allows the select (using a slider) different values of n, and then it outputs the sum of:

$$\sum_{k=1}^{n} k^2.$$

(12) Extend the previous Sage Interact so that is work with any function given by the user, not only $f(x) = x^2$.

(13) Create a Sage Interact that allows the enter a text and counts the number of vowels in it.

(14) Extend the previous Sage Interact so that the user can select whether to count the number of vowels or consonants, and it displays the count.

(15) Create a Sage Interact that displays a random basic algebra question and waits for the user's answer. Then it checks the answer and displays whether the answer is correct or wrong.

(16) Extend the previous Sage Interact so that the user can answer 10 such random questions. At the end, you should display the number of correct answers and the number of wrong answers.

(17) Extend the previous Sage Interact so that the user can use a selector to pick how many questions to be displayed. At the end, you should display the number of correct answers and the number of wrong answers. Also keep track of the number of empty answers.

(18) Create a Sage Interact that will accept a number in base 2, and it will output its corresponding value in base 10.

(19) Create a Sage Interact that will accept a number in base 4, and it will output its corresponding value in base 10.

(20) Create a Sage Interact that will accept a number in base 16, and it will output its corresponding value in base 10.

(21) Create a Sage Interact that will accept a number in base n (where $n \leq 16$), and it will output its corresponding value in base 10.

(22) Create a Sage Interact that will allow the user to use a accept a number in base 10, and it will output its corresponding value in base 16.

(23) Create a Sage Interact that will allow the user to use a accept a number in base 10, and it will output its corresponding value in base 8.

(24) Create a Sage Interact that will allow the user to use a accept a number in base 10, and it will output its corresponding value in base n (where $n \leq 16$).

3.7   APPLICATION TO DATA SECURITY: CAESAR'S CIPHER. INTERACTS, STRINGS, AND ENCRYPTION

The following Sage Interact will encrypt a given text using Caesar Cipher. The cipher changes each letter from a given text with the corresponding letter that is three positions down the alphabet and therefore is a **substitution cipher**. For example, 'a' would be replaced with a 'd', 'b' with a 'e', ..., 'y' with a 'b', and 'z' would get replaced by a 'c'.

Note: Only letters will be encrypted. This Sage Interact will leave all other characters such as digits, spaces, and question marks unchanged.

```
@interact
def CaesarInteract(text = input_box(default = "Alea iacta est", type = str)):
    encrypted = ""
    for pos in [0,1,..,len(text)-1]  :
        code = ord(text[pos])
        #is it a letter? then encrypt
        if( ( code>= ord('a') and code <=ord('w'))
                      or
            ( code>= ord('A') and code <=ord('W'))):
            code += 3

        #last three letters will cycle back!
        elif( ( code>= ord('x') and code <=ord('z'))
                      or
            ( code>= ord('X') and code <=ord('Z'))):
            code += 3-26

        encrypted +=chr(code)
    print encrypted
```

We obtain

text   Alea iacta estz

Dohd ldfwd hvwc

and, by changing the text value:

text   Sage Math is a great tool. We love it!

Vdjh Pdwk lv d juhdw wrro. Zh oryh lw!

In the next example, we modify our previous Sage Interact so it can allow the user to encrypt or decrypt a given message using Caesar Cipher. Also, we extended its capability so that it will allow the user to use a slider in order to select the value of the encryption/decryption key.

Note: To keep things simple, we completely redesigned the previous Sage Interact:

```
@interact
def CaesarInteract(text = input_box(default = "Alea iacta est", type = str),
                   to_do = selector(["Encrypt", "Decrypt"], default = "Encrypt"),
                   key = slider(0,25,1)):
    if to_do == "Encrypt":
        key = key
    else:
        key = -key

    encrypted = ""
    for pos in [0,1,..,len(text)-1]   :
        code = ord(text[pos])
        #is it a letter? then encrypt
        if ( code>= ord('a') and code <=ord('z')):
            code = ord('a')+ (code - ord('a')+key)%26
        elif( code>= ord('A') and code <=ord('Z')):
            code = ord('A')+ (code - ord('A')+key)%26
        encrypted +=chr(code)
    print encrypted
```

We obtain

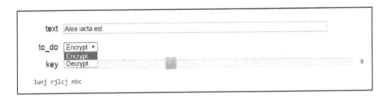

and

A similar cipher code is the Confederate Cipher Disk, which was used during the American Civil War. It consists of two concentric disks, each containing the 26 letters of the Latin alphabet written out clockwise. We challenge the reader to change the previous Sage Interact so that it implements a Confederate Cipher Disk.

3.7.1   Exercises

(1) Create a Sage Interact that implements the Confederate Cipher Disk. It should allow the user to select the letter of the inner circle that corresponds to letter 'A' in the outer circle.

(2) Create a Sage Interact that implements the Confederate Cipher Disk. It should allow the user to select a letter for the inner circle, and the corresponding letter in the outer circle.

(3) Design your own encryption/decryption algorithm and implement it using Sage Interacts.

3.8   APPLICATION TO BUSINESS: PRESENT VALUE OF AN ANNUITY. AMORTIZATION

The following Sage Interact computes the present value of an annuity that pays an amount *PMT* (*default* value: $250) per period, for *t* (*default* value: 6) years given that the money is worth *r* (*default* value: 6%) using compounded interest, with *m* periods a year (*default* value is: monthly), using the formula:

$$PV = PMT \frac{1 - \left(1 + \frac{r}{m}\right)^{-mt}}{\frac{r}{m}}$$

@interact

```
def PresentValueInteract(
    PMT = input_box(default = 250, label = 'periodic payment'),
    t = input_box(default = 6, label = "number of years"),
    r = input_box(default = 0.06, label = "annual rate"),
    m = selector(values = ["Monthly", "Quarterly", \
                    "Semi-annually", "Annually"], \
            default = "Monthly" , label = 'compounded')):

    if( m == "Monthly") :
        m = 12
    elif( m == "Quarterly") :
        m = 4
    elif( m == "Semi-annually") :
        m = 2
    elif( m == "Annually") :
        m = 1
    PV = PMT*(1-(1+r/m)^(-m*t))/(r/m)
    print "present value = ", PV
```

We obtain

periodic payment	250
number of years	6
annual rate	0.060000000000000
compounded	Monthly ▼

present value = 15084.8784838801

and

periodic payment	700
number of years	30
annual rate	0.05
compounded	Monthly ▼

present value = 130397.131932253

To output an amortization schedule, given a value for the present value (*default* $15084.88), the length of the loan t in years (*default* value: 6), and the annual interest rate r (*default* value: 6%) using compounded interest, with m periods a year (*default* value is: monthly), one could use the following Sage Interact:

```
@interact
def PresentValueInteract(
    PV = input_box(default = 15084.88, label = 'present value'),
    t = input_box(default = 6, label = "number of years"),
    r = input_box(default = 0.06, label = "annual rate"),
    m = selector(values = ["Monthly", "Quarterly", \
                           "Semi-annually", "Annually"], \
                 default = "Monthly" , label = 'compounded')):

    if( m == "Monthly") :
        m = 12
    elif( m == "Quarterly") :
        m = 4
    elif( m == "Semi-annually") :
        m = 2
    elif( m == "Annually") :
        m = 1

    PMT =PV*(r/m) / ((1-(1+r/m)^(-m*t)))
```

```
period = 0
print 'period\t payment amount \tprincipal part' \
        + '\t\tinterest paid \t\t balance'
while(PV>=0):
    period = period + 1
    interest_paid = PV*r/m
    principal_part = PMT - interest_paid
    PV -= principal_part
    print period, '\t',PMT, '\t', principal_part, \
                    '\t', interest_paid, '\t', PV
```

we obtain

present value	15084.8800000000
number of years	6
annual rate	0.060000000000000
compounded	Monthly ▼

period	payment amount	principal part	interest paid	balance
1	250.000025126485	174.575625126485	75.4244000000000	14910.3043748735
2	250.000025126485	175.448503252118	74.5515218743676	14734.8558716214
3	250.000025126485	176.325745768378	73.6742793581070	14558.5301258530
4	250.000025126485	177.207374497220	72.7926506292651	14381.3227513558
5	250.000025126485	178.093411369706	71.9066137567790	14203.2293399861
6	250.000025126485	178.983878426555	71.0161466999305	14024.2454615595
7	250.000025126485	179.878797818688	70.1212273077977	13844.3666637408
8	250.000025126485	180.778191807781	69.2218333187042	13663.5884719331
9	250.000025126485	181.682082766820	68.3179423596653	13481.9063891662
10	250.000025126485	182.590493180654	67.4095319458312	13299.3158959856
11	250.000025126485	183.503445646557	66.4965794799280	13115.8124503390
12	250.000025126485	184.420962874790	65.5790622516952	12931.3914874642
13	250.000025126485	185.343067689164	64.6569574373212	12746.0484197751
14	250.000025126485	186.269783027610	63.7302420988754	12559.7786367475
15	250.000025126485	187.201131942748	62.7988931837374	12372.5775048047
16	250.000025126485	188.137137602462	61.8628875240236	12184.4403672023
17	250.000025126485	189.077823290474	60.9222018360113	11995.3625439118
18	250.000025126485	190.023212406926	59.9768127195589	11805.3393315049
19	250.000025126485	190.973328468961	59.0266966575243	11614.3660030359
20	250.000025126485	191.928195111306	58.0718300151795	11422.4378079246
21	250.000025126485	192.887836086862	57.1121890396230	11229.5499718377
22	250.000025126485	193.852275267297	56.1477498591887	11035.6976965704
23	250.000025126485	194.821536643633	55.1784884828522	10840.8761599268
24	250.000025126485	195.795644326851	54.2043807996340	10645.0805156000
25	250.000025126485	196.774622548486	53.2254025779998	10448.3058930515
26	250.000025126485	197.758495661228	52.2415294652573	10250.5473973902
27	250.000025126485	198.747289130634	51.2527360860513	10051.8001002567

and

present value	2000
number of years	1
annual rate	0.06
compounded	Quarterly ▼

period	payment amount	principal part	interest paid	balance
1	518.889571976266	488.889571976266	30.0000000000000	1511.11042802373
2	518.889571976266	496.222915555910	22.6666564203560	1014.88751246782
3	518.889571976266	503.666259289249	15.2233126870174	511.221253178576
4	518.889571976266	511.221253178587	7.66831879767864	-1.14255271910224e-11

One can write some Sage function in order to format the output to two decimal places as shown below:

```
def format2Digits(xx):
    wholePart = int(xx)
    twoDigits = int( ((xx*100)-(wholePart*100)) )
    return   str(wholePart) + '.' + str(twoDigits)
def round2Digits(xx):
    return (int(xx*100+0.5))/100
@interact
def PresentValueInteract(
    PV = input_box(default = 15084.88, label = 'present value'),
    t = input_box(default = 6, label = "number of years"),
    r = input_box(default = 0.06, label = "annual rate"),
    m = selector(values = ["Monthly", "Quarterly", \
                        "Semi-annually", "Annually"], \
            default = "Monthly" , label = 'compounded')):

    if( m == "Monthly") :
        m = 12
    elif( m == "Quarterly") :
        m = 4
    elif( m == "Semi-annually") :
        m = 2
    elif( m == "Annually") :
        m = 1

    PMT = PV*(r/m) / ((1-(1+r/m)^(-m*t)))

    #rounding to two digits
    PMT =   round2Digits(PMT)

    period = 0
    print "period\tpayment amount \tprincipal part"        \
            +"\tinterest paid \tbalance"
    while(PV>0.01):
        period = period + 1
        interest_paid =   round2Digits( PV*r/m ).n()
        principal_part = PMT - interest_paid
        PV -= principal_part
        print period, '\t',format2Digits(PMT),                \
                    '\t\t', format2Digits(principal_part), \
                    '\t\t', format2Digits(interest_paid),  \
                    '\t\t', format2Digits(PV)
```

We get

present value	15084.8800000000			
number of years	6			
annual rate	0.0600000000000000			
compounded	Monthly ▼			

period	payment amount	principal part	interest paid	balance
1	250.0	174.58	75.42	14910.30
2	250.0	175.45	74.55	14734.84
3	250.0	176.33	73.67	14558.51
4	250.0	177.20	72.79	14381.31
5	250.0	178.9	71.91	14203.22
6	250.0	178.98	71.2	14024.24
7	250.0	179.88	70.12	13844.36
8	250.0	180.78	69.22	13663.58
9	250.0	181.68	68.31	13481.90
10	250.0	182.59	67.41	13299.31
11	250.0	183.50	66.50	13115.81
12	250.0	184.42	65.58	12931.39
13	250.0	185.34	64.66	12746.5
14	250.0	186.27	63.73	12559.78
15	250.0	187.20	62.80	12372.57
16	250.0	188.14	61.86	12184.43
17	250.0	189.8	60.92	11995.35
18	250.0	190.2	59.98	11805.33
19	250.0	190.97	59.3	11614.37
20	250.0	191.93	58.7	11422.43
21	250.0	192.89	57.11	11229.55
22	250.0	193.85	56.15	11035.70
23	250.0	194.82	55.18	10840.88
24	250.0	195.80	54.20	10645.8
25	250.0	196.77	53.23	10448.31
26	250.0	197.76	52.24	10250.54
27	250.0	198.75	51.25	10051.79
28	250.0	199.74	50.26	9852.6
29	250.0	200.74	49.26	9651.32

and

present value	2000			
number of years	1			
annual rate	0.0600000000000000			
compounded	Quarterly ▼			

period	payment amount	principal part	interest paid	balance
1	518.89	488.89	30.0	1511.11
2	518.89	496.22	22.67	1014.89
3	518.89	503.66	15.22	511.22
4	518.89	511.22	7.67	0.0

Update: one can also use the round method in order to round a value to include at most a specified amount of decimal places. For example, round(x,2) will round x to two decimal places (although it will not output the trailing 0s past the decimal point, if any). Using it, the previous code can be rewritten as

```
@interact
def PresentValueInteract(
    PV = input_box(default = 15084.88, label = 'present value'),
    t = input_box(default = 6, label = "number of years"),
    r = input_box(default = 0.06, label = "annual rate"),
    m = selector(values = ["Monthly", "Quarterly", \
                        "Semi-annually", "Annually"], \
                default = "Monthly" , label = 'compounded')):
    if( m == "Monthly") :
        m = 12
    elif( m == "Quarterly") :
        m = 4
    elif( m == "Semi-annually") :
        m = 2
    elif( m == "Annually") :
        m = 1
    PMT = PV*(r/m) / ((1-(1+r/m)^(-m*t)))
        #rounding to two digits
    PMT =  round(PMT , 2)
    period = 0
    print "period\tpayment amount \tprincipal part"        \
            +"\tinterest paid \tbalance"
    while(PV>0.01) :
        period = period + 1
        interest_paid =  round( PV*r/m , 2)
        principal_part = PMT - interest_paid
        PV -= principal_part
        print period, '\t',round(PMT, 2),                  \
                    '\t\t', round(principal_part, 2), \
                    '\t\t', round(interest_paid, 2),  \
                    '\t\t', round(PV, 2)
```

and obtain equivalent results.

3.8.1   Exercises

(1) Using a Finite Mathematics for Business book, create Sage Interacts that can be used to compute the future value of an ordinary annuity.

(2) Using a Finite Mathematics for Business book, create Sage Interacts that can be used to output a table of future value of an ordinary annuity.

3.9   APPLICATION TO ELEMENTARY STATISTICS: MEAN, MEDIAN, HISTOGRAMS, AND BAR CHARTS

One can make use of the mean, median, and mode, functions to obtain numerical description of a given set of data. Here is the first example:

```
@interact
def StatsInteract(
   data = input_box(default = [1, 2, 3, 3]),
   statistic = selector(values = ["mean", "median", "mode"],
                        label = "choose a numerical description:",
                        default = "mean")):

   if (statistic == "mean"):
      print mean(data)
   elif(statistic == "median"):
      print median(data)
   else:
      print mode(data)
```

We obtain

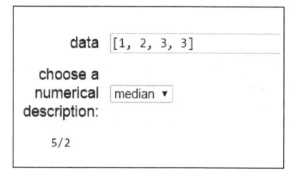

and

One can also give the `data` as a for loop as shown below:

data	`[sin(x) for x in [0,pi/12,..,pi]]`
choose a numerical description:	median ▾
	`1/2*sqrt(2)`

To obtain a **frequency histogram** using a given set of data, one can use the following Sage Interact:

```
import matplotlib.pyplot as plot

@interact
def HistogramInteract(
    data = input_box(default = [1, 2, 3, 3]),
    bins = slider(vmin=1, vmax = 10, step_size = 1) ):

    a = min(data[i] for i in [0,1,..,len(data)-1])
    b = max(data[i] for i in [0,1,..,len(data)-1])

    plot.hist(data, bins, range=(a,b))
    plot.show()
    plot.close()
```

and obtain

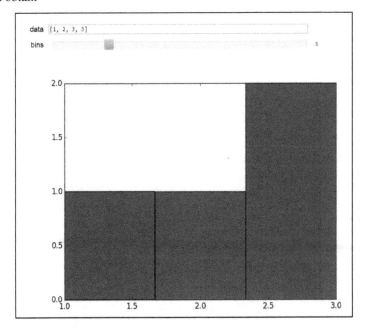

and

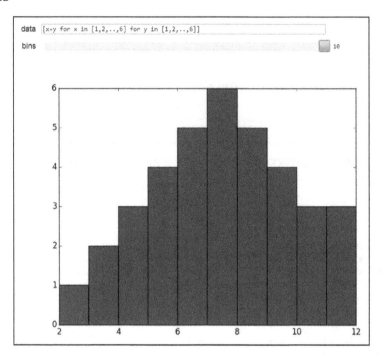

To obtain scatter plots, one can use Sage Interacts as the one shown below:

```
@interact
def HistogramInteract(
    data = input_box(default = [(1, 2), (3, 3), (3,5), (5, 7), (8,9)])):

    a = min(data[i] for i in [0,1,..,len(data)-1])
    b = max(data[i] for i in [0,1,..,len(data)-1])

    plot = scatter_plot(data,
                        figsize=4,
                        facecolor="yellow",
                        edgecolor="blue",
                        markersize=30,
                        marker='s')
    plot.show()
```

to obtain

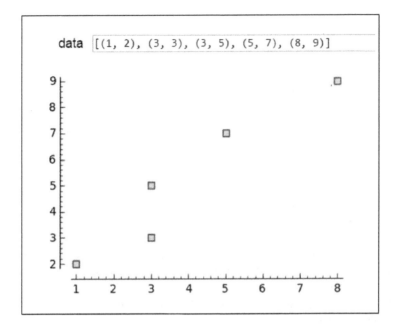

and

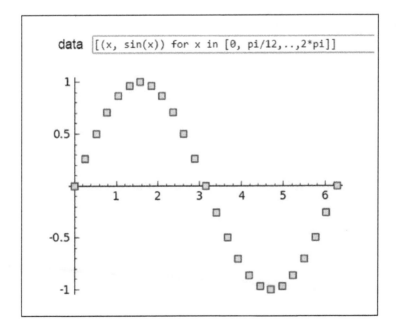

If you want to plot something else instead of squares, then you can change the option marker = 's' with any of the following options:

- 'o' (for circle)
- 'D' (for diamond)
- 'H' (for hexagon)
- '_' (for horizontal line)
- 'p' (for pentagon)
- '.' (for point)
- 'v' (for triangle down)
- '<' (for triangle left)
- '>' (for triangle right)
- '^' (for triangle up)
- '|' (for vertical line)

One could also write Sage code to implement all the statistical functions that compute the mean, mode, range, and so on instead of using library functions. We challenge the user to attempt it.

The following Sage Interact can be used to graph **Bar Charts**. In this interact, we randomly generate a set of default values, then we create the corresponding Bar Chart.

```
import random
@interact
def HistogramInteract(
    data = input_box(default = [random.randint(1,20) for x in [1,2,..,7]])):

    plot = bar_chart(data, color="red")
    plot.show()
```

We obtain

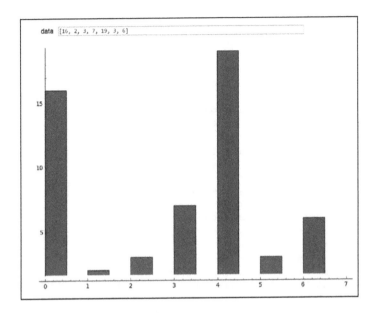

To create your own code that computes the mean and variance for a given set of data points, you can implement a Sage Interact similar to the one given next:

```
var('x,y')
@interact
def MeanVarianceInteract(
        data = input_box( default = [5, 8, 7, 6, 9]),
        dataType = selector(values = ["Population", "Sample"],
                            default = "Population")):
```

```
if(len(data)>1):
    #computing the mean
    sum = 0
    for position in [0,1,..,len(data)-1]:
            sum += data[position]
    mean = sum/len(data)

    #computing the variance
    variance = 0
    for i in [0,1,..,len(data)-1]:
            variance += (data[i]-mean)^2

    if (dataType == "Population"):
            variance /= len(data)
    else:
            variance /= len(data)-1
    #computing the std deviation
    std_dev = sqrt(variance)

    #output the values
    print "mean = ",  mean
    print "variance = ",  variance
    print "standard deviation = ",  std_dev

else:
    print "not enough data!"
```

We obtained

and

Note: Using the sum function one could equivalently rewrite the previous Sage code to

```
var('x,y')
@interact
def MeanVarianceInteract(
        data = input_box( default = [5, 8, 7, 6, 9]),
        dataType = selector(values = ["Population", "Sample"],
                            default = "Population")):

    if(len(data)>1):

        #computing the mean
        #to avoid compiling error/confusion sum was renamed to summ
        summ = sum([data[position] for position in [0,1,..,len(data)-1]])
        mean = summ/len(data)

        #computing the variance
        variance = sum([(data[i]-mean)^2 for i in [0,1,..,len(data)-1]])

        if (dataType == "Population"):
                variance /= len(data)
        else:
                variance /= len(data)-1
        #computing the std deviation
        std_dev = sqrt(variance)

        #output the values
        print "mean = ",   mean
        print "variance = ",   variance
        print "standard deviation = ",   std_dev
    else:
        print "not enough data!"
```

3.9.1 Exercises

(1) Using a Statistics book as a guide, implement a Sage Interact that can be used to check whether a given data set is unimodal, bimodal, or multimodal. If applicable, compute the mode. The Interact should allow the user to input a data set.

(2) Using a Statistics book as a guide, implement a Sage Interact that can be used to compute a frequency table for a given data set. The Interact should allow the user to input a data set. Hint: You may want to sort the data set before computing the frequency of each table. Then update the Sage Interact just produced, so the user can use a selector to choose whether to output the absolute frequencies or relative frequencies.

(3) Using a Statistics book as a guide, implement a Sage Interact that can be used to compute the range of a given data set.

(4) Using a Statistics book as a guide, implement a Sage Interact that computes the five-number summary of a given data set.

(5) Using a Statistics book as a guide, implement a Sage Interact that allows the user to input a data set, and use a selector to choose a value x between 1 and 100. Then, it will output the xth percentile.

(6) Using a Statistics book as a guide, implement a Sage Interact that allows the user to input a data set, and a value x. Then, it will output the percentile rank of that value.

(7) Using a Statistics book as a guide, implement a Sage Interact that allows the user to input a data set, and use a selector to choose a value x between 1 and 100. Then, it will output the $x\%$ confidence interval for estimating the population mean.

(8) Using a Statistics book as a guide, implement a Sage Interact that allows the user to input a data set, and use a selector to choose a value x between 1 and 100. Then, it will output the $x\%$ confidence interval for estimating the population variance.

4

SAGE INTERACTS FOR NUMERICAL METHODS

In this chapter, we see several examples on how one can create Sage Interacts for various Numerical Methods. We either implement some of the algorithms completely from scratch or make use of the library functions.

Note: Please be aware that in here you will be expected to either master the Mathematical topics concepts considered, or to bundle this book with specialized books that cover them. Some recommended books that cover the Mathematical topics considered here are [2], [5; 7; 17; 8; 18].

Note: In order to fully understand these Sage Interacts you will be expected to master most of the previous sections that cover the basic Sage. If needed, take your time to review all of them before proceeding with this chapter.

4.1 EQUATIONS OF LINES

To get you up to speed, we start this chapter with a simple interactive example. Here we create a Sage Interact that allows the user to use two sliders: m, b and then it will plot the function $y = mx + b$.

```
@interact
def Lines(m = slider(-10, 10, 1),
          b = slider(-10, 10, 1, 0)):
    plot(m*x+b, -10,10, title = "y = "+ str(m*x+b)).show()
```

An Introduction to SAGE Programming: With Applications to SAGE Interacts for Numerical Methods, First Edition. Razvan A. Mezei

We obtain

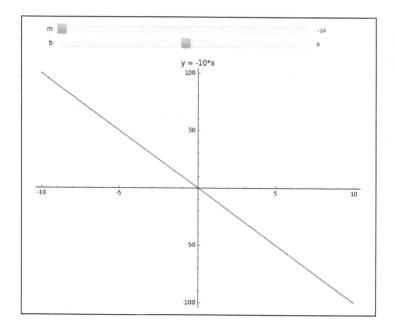

and

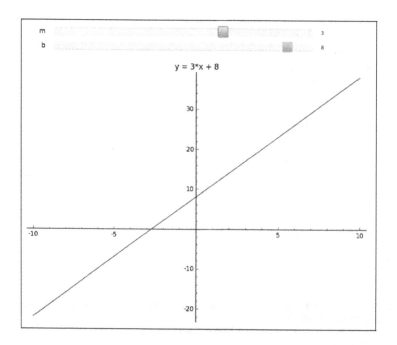

Note: Instead of using m = `slider(vmin = -10,vmax = 10, step_size = 1)`, we used the simplified m = `slider( -10, 10, 1)`. One can supply a fourth value to the `slider`, in which case that value will be used as a *default* value of the *slider*.

4.1.1 Exercises

(1) Create a Sage Interact that will draw a parabola $y = ax^2 + bx + c$, and will allow the user to use sliders for each coefficient a, b, c.

(2) Improve the previously created interact so that it allows the user to specify the drawing color and the thickness of the graph using selectors.

(3) Create a Sage Interact that will draw $y = a^x$, and will allow the user to use a slider for the constant a.

(4) Improve the previously created interact so that it allows the user to specify the drawing color and the thickness of the graph using selectors.

(5) Create a Sage Interact that will draw $y = a \cdot \cos(bx + c)$, and will allow the user to use a slider for each of the constants a, b, c.

(6) Improve the previously created interact so that it allows the user to specify the drawing color and the thickness of the graph using selectors.

(7) Create a Sage Interact that will allow the user to input any function $f(x)$, and specify a domain $[a, b]$, then it will graph the function.

(8) Improve the previously created interact so that it allows the user to specify the drawing color and the thickness of the graph using selectors.

4.2 TANGENT LINES AND PLOTS

One can use plots in conjunction with interacts and obtain really neat applications. Next we create a Sage Interact that can be used to graph $f(x) = \sin(x)$ and the tangent line to the graph of the function at $x = x_0$. We use a *slider* to select the value of the point x_0, and use a *selector* to allow the user to pick different colors for the tangent line.

```
@interact
def TangentInteract(
    x0 = slider(vmin=-pi, vmax=pi, step_size=pi/12,
            default=0, label="Select the point x0"),
            tancolor = selector(values = ["red", "blue",\
                    "yellow", "purple", "green"],
            label = "Tangent line's color: ",default = "red" )):
    #here comes the body of the interact:
    #first we plot the function:
    p1 = plot(sin(x), x,-pi,pi)

    #then we compute and plot the tangent line at x0
    df(x) = diff(sin(x))
    m = df(x0)
    p2 = plot( m*(x-x0)+sin(x0) , x,-pi,pi, color = tancolor)
```

```
#for a better view,
#we plot the tangent point
p3 = list_plot([(x0, sin(x0))], size = 50, color=tancolor)

(p1+p2+p3).show(title = "y = "+ str(sin(x)))
```

We obtained:

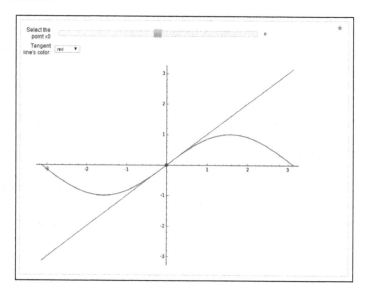

Changing the point x_0 to $\frac{5\pi}{12}$, and the tangent line's color to "green," we obtain

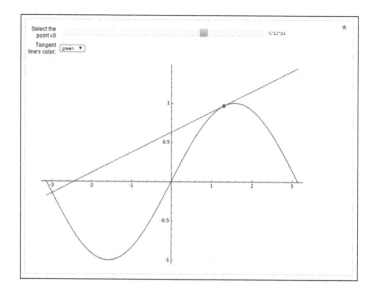

To change the previous Sage Interact so that it allows the user to use any function, not only $f(x) = \sin(x)$, one can use

```
@interact
def TangentInteract2(
    f = input_box(default=sin(x) ),
    x0 = slider(vmin=-pi, vmax=pi, step_size=pi/12,
                default=0, label="Select the point x0"),
    tancolor = selector(values = ["red", "blue", "yellow",
    "purple", "green"],
                label = "Tangent line's color: ",default = "red" )):
    #here comes the body of the interact:
    #first we plot the function:
    p1 = plot(f(x), x,-pi,pi)

    #then we compute and plot the tangent line at x0
    df(x) = diff(f(x))
    m = df(x0)
    p2 = plot( m*(x-x0)+f(x0) , x,-pi,pi, color = tancolor)

    #for a better view,
    #we plot the tangent point
    p3 = list_plot([(x0, f(x0))], size = 50, color=tancolor)

    (p1+p2+p3).show()
```

and obtain

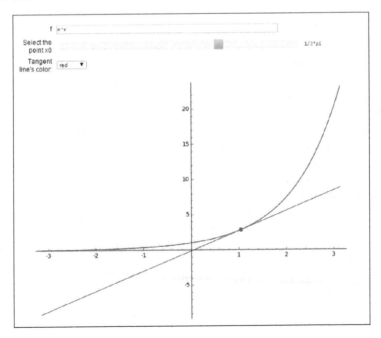

and

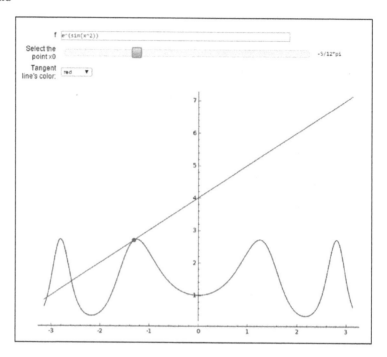

4.2.1 Exercises

(1) Change the Sage Interacts given in this section so that it graphs the function
$f(x) = \sin(x)$, and its derivative. Using a slider the user will be able to choose
a point x_0 from the given interval, and the interact will display the value of the
derivative at that point x_0. Explain the connection between the value (sign) of the
derivative at the point x_0 and the monotonicity of the function $f(x)$ around x_0.

(2) Change the Sage Interact obtained in the previous exercise so that the user can
enter any desired function in place of $f(x) = \sin(x)$.

(3) Change the Sage Interacts given in this section so that it graphs the function
$f(x) = \sin(x)$, and its second derivative. Using a slider the user will be able to
choose a point x_0 from the given interval, and the interact will display the value
of the first and second derivatives at that point x_0. Explain the connection between
the value of the second derivative of f around the point x_0, the value of the first
derivative of the function $f(x)$ around x_0, and the value of the original function f
around x_0.

(4) Change the Sage Interact obtained in the previous exercise so that the user can
enter any desired function in place of $f(x) = \sin(x)$.

4.3   TAYLOR POLYNOMIALS

The following Sage Interact will allow the user to input a function f, and a domain $[a, b]$. Then, it will compute and plot the Taylor polynomial of a degree n selected by the user (by means of a slider), around a specified point (around), and also plot the original function f. Since we have so many parameters, it is recommended to use an "Update" button so the graph only gets updates once we finish inputting all the desired changes:

```
@interact
def TaylorPlotInteract(
    f = input_box(default = e^x),
    a = input_box(default = 0),
    b = input_box(default = 3),
    around  = input_box(default = 0),
    n = slider(vmin=0, vmax=10,\
               step_size=1, default=3),
    auto_update=False):

    f(x) = f
    p = plot(f(x), x, a, b,
             color = 'red',
             legend_label = "$f(x)  = "+str(f(x))+"$" )
    t(x)=f(x).taylor(x,around,n)
    p += plot(t(x), x, a, b,
              color = 'blue',
              legend_label = "$p_"+str(n)+"(x)="+str(t(x))+"$")
    p.show()
```

We obtain:

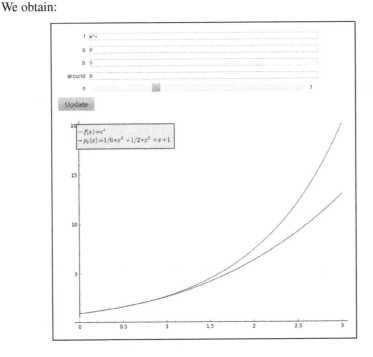

and

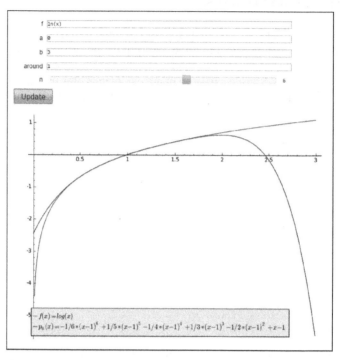

For the function $\cos(x)$ on $[-\pi, \pi]$, we obtain

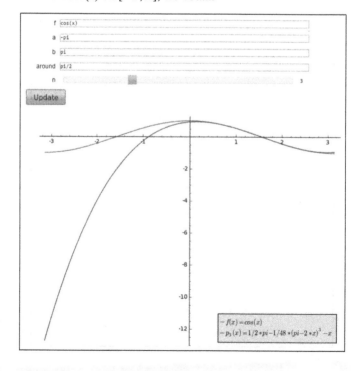

One can modify the Sage Interact given above so it allows the user to input a set of values for *n*, and then it plots the function $f(x)$ along with the Taylor polynomials of each degree given:

```
@interact
def TaylorPlotInteract(
    f = input_box(default = e^x),
    a = input_box(default = 0),
    b = input_box(default = 3),
    around  = input_box(default = 0),
    degrees = input_box(default  = [1 , 2]),
    auto_update=False):

    f(x) = f
    p = plot(f(x), x, a, b,
            color = 'red',
            legend_label = "$f(x) = "+str(f(x))+"$" )

    for n in degrees:
        t(x)=f(x).taylor(x,around,n)
        p += plot(t(x), x, a, b,
            color = 'blue',
            legend_label = "$p_"+str(n)+"(x)="+str(t(x))+"$")
    p.show()
```

We obtain

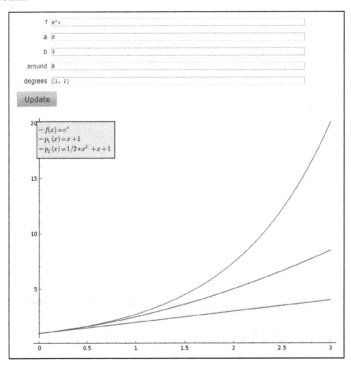

and by changing the array of degrees to [1, 2, 5, 7] we obtain

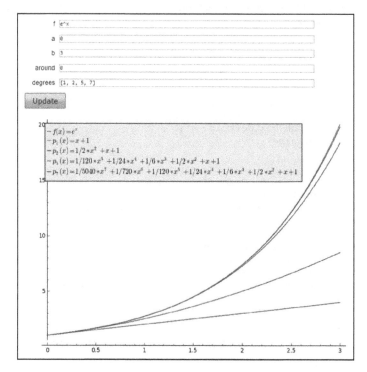

To obtain individualized colors for each of the Taylor polynomials, one can replace the `color` option with `rgbcolor=hue(1/n)`:

```
@interact
def TaylorPlotInteract(
    f = input_box(default = e^x),
    a = input_box(default = 0),
    b = input_box(default = 3),
    around  = input_box(default = 0),
    degrees = input_box(default  = [1 , 2]),
    auto_update=False):

    f(x) = f
    p = plot(f(x), x, a, b,
             color = 'black',
             legend_label = "$f(x)  = "+str(f(x))+"$" )

    for n in degrees:
        t(x)=f(x).taylor(x,around,n)
        p += plot(t(x), x, a, b,
             rgbcolor=hue(1/n) ,
             legend_label = "$p_"+str(n)+"(x)="+str(t(x))+"$")
    p.show()
```

and obtain:

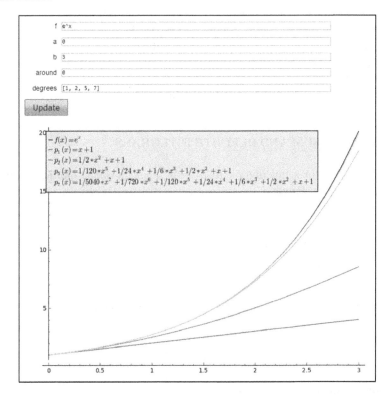

4.3.1 Exercises

(1) Change the aforementioned interact so that it outputs all the Taylor polynomials of degree up to n, where n is selected by the user through a slider (n between 0 and 10 inclusively).

(2) Change the aforementioned interact so that it allows the user to select the colors for the graphs of f and the Taylor polynomials, and choose the thickness in each case.

(3) Simplify the interact given in this section so that it uses a fixed function, $f(x) = \cos(x)$, and a fixed interval $[a, b]$ (let's say $[-\pi, \pi]$), and it only allows the user to select the value of the degree n, and the value of *around*.

(4) Simplify the interact given in this section so that it uses a fixed function, $f(x) = e^x$, and a fixed interval $[a, b]$ (let's say $[-1, 3]$), and it only allows the user to select the value of the degree n, and the value of *around*.

(5) Create a Sage Interact that will find an approximation for e using Maclaurin polynomials of degree at most n (selected by the user). The interact will display the approximation, as well as an approximate value of the error of approximation.

(6) Modify the interact found in the previous exercise so that it will use Taylor polynomials of degree at most n, around some point specified by the user.

(7) Create a Sage Interact that will find an approximation for ln 2 using Maclaurin polynomials of degree at most n (selected by the user). The interact will display the approximation, as well as an approximate value of the error of approximation.

(8) Modify the interact found in the previous exercise so that it will use Taylor polynomials of degree at most n, around some point specified by the user.

4.4   RIEMANN SUM AND DEFINITE INTEGRALS

Recall ([17]) that for a function $f : [a, b] \to \mathbb{R}$, one can divide the interval into n subintervals of equal length

$$\Delta x = \frac{b-a}{n},$$

and consider the right endpoints for each subinterval:

$$x_i = a + i \cdot \Delta x, \; i = 1, 2, \ldots, n.$$

Then, the following is a Riemann Sum (using right endpoints):

$$S_n = \sum_{i=1}^{n} f(x_i)\Delta x.$$

The **definite integral** is defined as

$$\int_a^b f(x)dx = \lim_{n \to \infty} \sum_{i=1}^{n} f(x_i)\Delta x$$

provided the limit exists.

We next give a Sage Interact that allows the user to enter a function f and an interval $[a, b]$. It then computes the Riemann Sum (using right endpoints!) and by taking the limit as $n \to \infty$, it also computes the definite integral of the given function.

```
var('i,n')
@interact
def RiemannInteract(
    f = input_box(default = e^x),
    a = input_box( default = 0),
    b = input_box( default = 1)):

    f(x) = f
    dx = (b-a)/n

    RiemannSum = sum( f(a+i*dx)*dx,    i, 1, n)
    DefInt = lim(RiemannSum, n=infinity)
```

```
print "Riemann Sum: "
show(RiemannSum)
print "Definite Integral: "
show(DefInt )
```

We obtain:

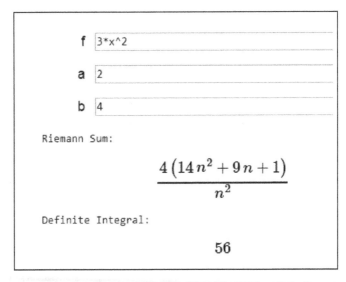

and, by changing the function $f(x)$ to $3x^2$ and the limits of integration we get:

The following Sage Interact will graph the strips (rectangles) that appear in the computation of the Riemann Sum (using right endpoints!) for a given function $f(x)$, on a given interval $[a, b]$. The user will use a slider to select n the number of strips to be used.

```
@interact
def RiemannSum(
    f = input_box(default = e^x) ,
    a = input_box(default =  0) ,
    b = input_box(default = 1) ,
    n = slider(vmin = 1, vmax = 20, default = 4, step_size=1)  ):
    f(x)=f
    #plot the function
    p = plot(f, x, a,b, color = 'red', thickness = 4)
    dx = (b-a)/n
    #plot each trapezoids
    for i in [1,2,..,n]:
        #for the shading ...
        xbegin = a+(i-1)*dx
        xend   = a+i*dx

        p += polygon( [(xbegin,0),   (xend,0),                \
                       (xend, f(xend)), (xbegin,f(xend))],   \
                       color='yellow',aspect_ratio='automatic')
        #these are optional for a better graph:
        p += line([(xend,0), (xend, f(xend))], color = 'black')
        p += line([(xend, f(xend)), (xbegin,f(xend))], color = 'black')
        p += line([(xbegin,f(xend)),(xbegin,0) ], color = 'black')
    p.show()
```

We get

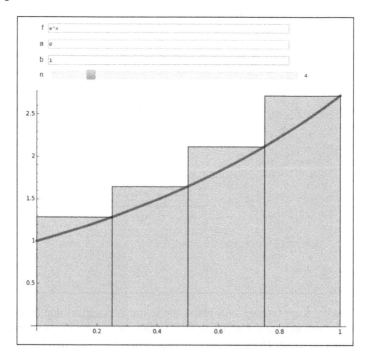

and

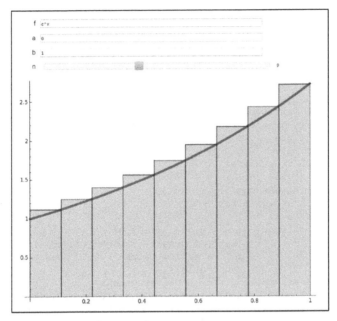

If we make use of options `title.` and `ticks`, that is replacing the last line in the previous code with

```
p.show(title = 'Riemann Sum with Right Endpoints', ticks = dx)
```

we obtain:

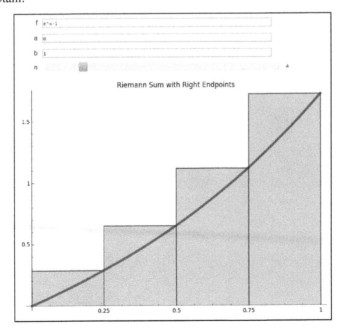

We can also use the option `alpha` in order to select the opacity of the filling color:

```
@interact
def RiemannSum(
    f = input_box(default = sin(x)) ,
    a = input_box(default =  0) ,
    b = input_box(default = 2*pi) ,
    n = slider(vmin = 1, vmax = 20, default = 8, step_size=1)  ):
    f(x)=f
    #plot the function
    p = plot(f, x, a,b, color = 'red', thickness = 4, \
            alpha = 0.5, fillcolor = 'blue', fill = true)
    dx = (b-a)/n
    #plot each trapezoids
    for i in [1,2,..,n]:
        #for the shading ...
        xbegin = a+(i-1)*dx
        xend   = a+i*dx
        p += polygon( [(xbegin,0),   (xend,0),                     \
                       (xend, f(xend)), (xbegin,f(xend))], \
                       color='yellow',alpha = 0.5, aspect_ratio=
                       'automatic')
        #these are optional for a better graph:
        p += line([(xend,0), (xend, f(xend))], color = 'black')
        p += line([(xend, f(xend)), (xbegin,f(xend))], color = 'black')
        p += line([(xbegin, f(xend)),(xbegin,0) ], color = 'black')
    p.show(title = 'Riemann Sum with Right Endpoints', ticks = dx)
```

We obtain:

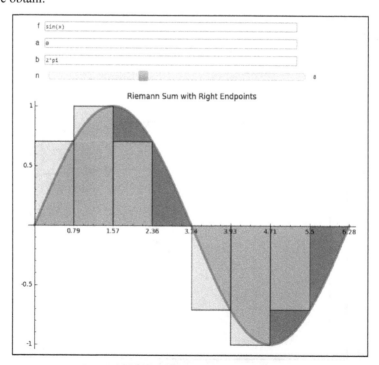

4.4.1 Exercises

(1) Using a Calculus book as a guide, create a Sage Interact that computes the value of the Riemann Sum of a given function and given interval using the midpoint rule.

(2) Using a Calculus book as a guide, create a Sage Interact that computes the value of the Riemann Sum of a given function and given interval using left endpoints for each subinterval.

(3) Using a Calculus book as a guide, create a Sage Interact that computes the value of the Riemann Sum of a given function and given interval using random points from each subinterval.

(4) Using a Calculus book as a guide, create a Sage Interact that computes the value of the Riemann Sum of a given function and given interval using a set of points given by the user.

(5) Using a Calculus book as a guide, create a Sage Interact that computes the value of the Riemann Sum of a given function and given interval using the left endpoints and the right endpoints rules.

(6) Using a Calculus book as a guide, create a Sage Interact that graphs the Riemann Sum of a given function and given interval using the midpoint rule.

(7) Using a Calculus book as a guide, create a Sage Interact that graphs the Riemann Sum of a given function and given interval using left endpoints for each subinterval.

(8) Using a Calculus book as a guide, create a Sage Interact that graphs the Riemann Sum of a given function and given interval using random points from each subinterval.

(9) Using a Calculus book as a guide, create a Sage Interact that graphs the Riemann Sum of a given function and given interval using a set of points given by the user.

(10) Using a Calculus book as a guide, create a Sage Interact that graphs the Riemann Sum of a given function and given interval using the left endpoints and the right endpoints rules.

4.5 TRAPEZOIDAL RULE FOR NUMERICAL INTEGRATION

There are cases when one cannot compute exactly the value of $\int_a^b f(x)dx$ even if it exists. Given a function $f(x)$, one can approximate $\int_a^b f(x)dx$ using the following **Trapezoidal Rule with evenly spaced nodes**:

$$\int_a^b f(x)dx \simeq h\left[\frac{1}{2}f(x_0) + f(x_1) + f(x_2) + \cdots + f(x_{n-1}) + \frac{1}{2}f(x_n)\right],$$

where

$$h = \frac{b-a}{n}$$

is the length of each subinterval, and

$$x_i = a + i * h, \quad i = 0, 1, \dots, n$$

are *evenly spaced nodes*. See [2] or [7] for more details about this topic.

We give below a Sage Interact that can be used for a graphical representation of the Trapezoidal Rule:

```
@interact
def TrapezoidalInteract(
    f = input_box(default = e^x) ,
    a = input_box(default =   0) ,
    b = input_box(default = 4) ,
    n = slider(vmin = 1, vmax = 30, default = 4, step_size=1)  ):
    f(x)=f
    #plot the function
    p = plot(f, x, a,b)

    h = (b-a)/n
    #plot each trapezoids
    p += line( [(a,0),  (a,f(a))] , color = "black")
    for i in [1,2,..,n]:
        p += line( [(a+(i-1)*h,f(a+(i-1)*h)), (a+i*h,f(a+i*h))] , \
                    color = "black")
        p += line( [(a+i*h,0),  (a+i*h,f(a+i*h))] ,                \
                    color = "black")

    p.show()
```

We obtain

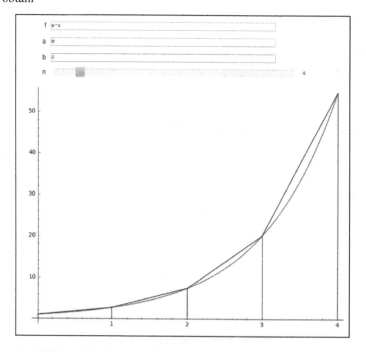

and

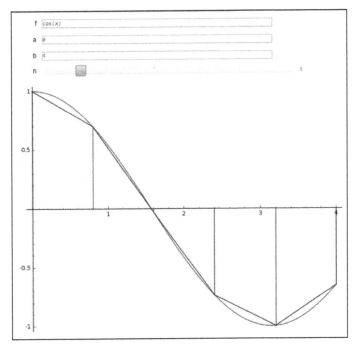

If you want to include a shading for the Trapezoidal Rule, then you can include a polygon and specify a shade color like in the code below:

```
@interact
def TrapezoidalInteract(
    f = input_box(default = e^x) ,
    a = input_box(default =  0) ,
    b = input_box(default = 4) ,
    n = slider(vmin = 1, vmax = 30, default = 4, step_size=1)  ):

    f(x) =f

    #plot the function
    p = plot(f, x, a,b, color = 'red', thickness = 4)

    h = (b-a)/n

    #plot each trapezoids
    p += line( [(a,0),   (a,f(a))] , color = "black")
    for i in [1,2,..,n]:
        p += line( [(a+(i-1)*h,f(a+(i-1)*h)), (a+i*h,f(a+i*h))] , \
                   color = "black")
        p += line( [(a+i*h,0),   (a+i*h,f(a+i*h))] , \
                   color = "black")

        #for the shading ...
        p += polygon( [(a+(i-1)*h,0),                  \
                       (a+(i-1)*h,f(a+(i-1)*h)),       \
```

```
                (a    +i*h, f(a+i*h)),              \
                (a+    i*h,           0)],           \
                color='yellow',                     \
                aspect_ratio='automatic')
    p.show()
```

and obtain

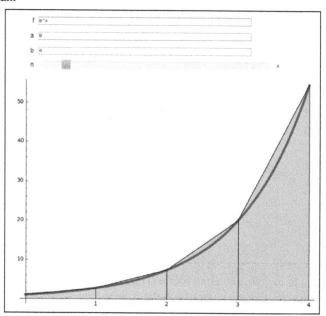

and

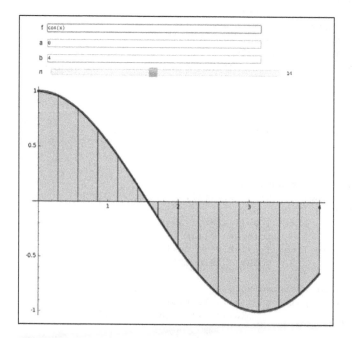

To obtain an approximation of the integral $\int_a^b f(x)dx$ using the Trapezoidal Rule, we can use the following Sage Interact:

```
def Trapezoidal(f, a, b, n):
    h = (b-a)/n

    S = 1/2*f(a)
    for i in [1,2,..,n-1] :
        S += f(a+i*h)
    S += 1/2*f(b)
    S *= h

    return S.n()

@interact
def TrapezoidalInteract(
    f = input_box(default = e^x) ,
    a = input_box(default =  0) ,
    b = input_box(default = 10) ,
    n = input_box(default =  8)  ):

    f(x)=f   ·
    print  "Approximation =",  Trapezoidal(f, a, b, n)
```

We obtain

```
    f  e^x

    a  0

    b  10

    n  8

Approximation = 24821.3541532970
```

The previous Sage Interact can be improved by allowing the user to select a value of n through a slider:

```
def Trapezoidal(f, a, b, n):
    h = (b-a)/n

    S = 1/2*f(a)
    for i in [1,2,..,n-1] :
        S += f(a+i*h)
    S += 1/2*f(b)
    S *= h

    return S.n()
@interact
def TrapezoidalInteract(
    f = input_box(default = e^x) ,
```

```
    a = input_box(default =   0) ,
    b = input_box(default = 10) ,
    n = slider(vmin = 1, vmax = 30, default = 4, step_size=1)  ) :

    f(x)=f
    print  "Approximation =",  Trapezoidal(f, a, b, n)
```

We obtain

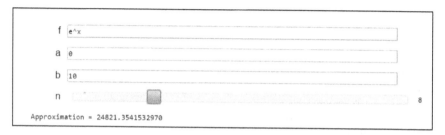

One can use the NumPy package to compute an approximate value of a definite integral using the Trapezoidal Rule. For more details on this, the reader may want to refer to the official website for it: [10]. In the code given below, we need to import the NumPy package, compute the intermediate points $a = x_0, x_1, \ldots, x_n = b$, and their corresponding y values (using $y = f(x)$). Then, we simply call the NumPy method $\texttt{trapz}$ as shown below:

```
import numpy
@interact
def TrapezoidByNumPy(
    f = input_box(default = e^x),
    a = input_box(default = 0),
    b = input_box(default = 3),
    n = slider(vmin=2, vmax=50, step_size = 1)):

    #needed to avoid the warning message
    f(x)=f

    #compute the size of each subinterval
    delta_x = (b-a)/n

    #compute the intermediate points
    xi_values = [a,a+delta_x,..,b]

    #compute the corresponding y values
    fxi_values = [f(i) for i in xi_values]

    #compute the Trapezoidal Rule value
    print "approximation = ",  numpy.trapz(fxi_values ,
    dx=delta_x).n(), \
        "\t exact value = ",  integrate(f(x), x, a, b).n()
```

For comparison purposes, we also output the value of the `integrate` method:

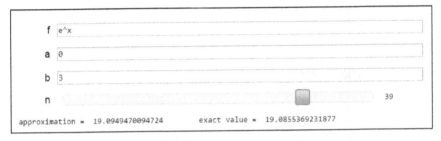

and

```
f  e^x
a  0
b  3
n                                          39
approximation = 19.0949470094724   exact value = 19.0855369231877
```

Note: One can also use the SciPy package for an alternative method to compute an approximate value of a definite integral using Trapezoidal Rule. For this, look for `scipy.integrate.trapz`.

4.5.1 Exercises

(1) Using a Numerical Analysis book as a guide, implement a Sage Interact for computing an approximation of $\int_a^b f(x)dx$ using Simpson's Rule.

(2) Using a Numerical Analysis book as a guide, implement a Sage Interact for plotting a graphical representation of approximating $\int_a^b f(x)dx$ using Simpson's Rule.

(3) Using a Numerical Analysis book as a guide, implement a Sage Interact for computing an approximation of $\int_a^b f(x)dx$ using Romberg Method.

(4) Using a Numerical Analysis book as a guide, implement a Sage Interact for computing an approximation of $\int_a^b f(x)dx$ using various Gaussian Quadratures.

(5) Using a Numerical Analysis book as a guide, implement a Sage Interact for computing an approximation of $\int_a^b f(x)dx$ using various Closed Newton–Cotes.

(6) Using a Numerical Analysis book as a guide, implement a Sage Interact for computing an approximation of $\int_a^b f(x)dx$ using various Open Newton–Cotes.

(7) Using a Numerical Analysis book as a guide, implement a Sage Interact for solving initial-value problems (IVPs) using Runge–Kutta Method of Order 2.

(8) Using a Numerical Analysis book as a guide, implement a Sage Interact for solving IVPs using Runge–Kutta Method of Order 4.

(9) Change the aforementioned Sage Interact so that they use `scipy.integrate.trapz` from SciPy package.

(10) Find the corresponding method from the SciPy package that implements Simpson's Rule and then use it in the Sage Interacts that were implemented in the exercises given above.

(11) Find the corresponding method from the SciPy package that implements the Romberg Method and then use it in the Sage Interacts that were implemented in the exercises given above.

(12) Find the corresponding method from the SciPy package that implements the Gaussian Quadratures and then use it in the Sage Interacts that were implemented in the exercises given above.

4.6 BISECTION ALGORITHM FOR SOLVING EQUATIONS

Let f be a continuous function on the closed interval $[a, b]$ such that $f(a) * f(b) < 0$. Then, one can find an approximate (or exact) solution x_0 to the equation $f(x) = 0$, using the following algorithm, known as the **Bisection Algorithm**:

- For the interval $[a, b]$ we investigate its midpoint, $\dfrac{a+b}{2}$. Exactly one of the following holds:

  - If $f\left(\dfrac{a+b}{2}\right) = 0$, then we are done. We found an exact solution, $x_0 = \dfrac{a+b}{2}$.

  - If $f(a) * f\left(\dfrac{a+b}{2}\right) < 0$, we repeat this process and search for a solution, in the first half of the interval $[a, b]$, that is the interval $\left[a, \dfrac{a+b}{2}\right]$.

  - If $f\left(\dfrac{a+b}{2}\right) * f(b) < 0$, we repeat this process and search for a solution, in the second half of the interval $[a, b]$, that is the interval $\left[\dfrac{a+b}{2}, b\right]$.

- We keep repeating these iterations until we either find an exact solution, or the length of the interval is small enough that any value from it could be considered a good approximation of the exact solution.

See [2] or [7] for more details about this topic.

Here is the Sage Interact that implements the previous algorithm:

```
@interact
def BisectionInteract(
    f = input_box(default = x^2-7, label = 'f(x)' ),
    a = input_box(default = 0 ),
    b = input_box(default = 7 ),
    tolerance = input_box(default= 0.00001) ):
```

```
x0 = (a+b)/2
while( f(x0)!=0 and   (b-a) >= tolerance):
    if(f(a)*f(x0)<0):
        b = x0
    else:               #(f(b)*f(x0)<0)
        a = x0
    x0 = (a+b)/2
print "approximate solution = ", x0.n()
```

We obtain the output:

```
approximate solution = 2.64575147628784
```

Changing the `tolerance` level to `0.001`, we obtain

```
approximate solution = 2.64593505859375
```

Changing the `tolerance` level to `0.00000001`, we obtain

```
approximate solution = 2.64575131004676
```

f(x)	x^2 - 7
a	0
b	7
tollerance	0.00000001

   approximate solution = 2.64575131004676

Compare this output to the output of the following Sage code:

`sqrt(7).n()`

which yields

```
2.64575131106459
```

Changing the function to $\cos(x)$, on $[0, \pi]$, as in

f(x)	cos(x)
a	0
b	pi
tollerance	0.00000001

   approximate solution = 1.57079632679490

we obtain

```
approximate solution = 1.57079632679490
```

which is essentially the same as the decimal representation of $\frac{\pi}{2}$: 1.5707963267
9490.

Note: To avoid the warning message

*Deprecation Warning: Substitution using function-call syntax and unnamed arguments
is deprecated and will be removed from a future release of Sage; you can use named
arguments instead, like EXPR(x=..., y=...)*

See http://trac.sagemath.org/5930 for details.

returned = f(*args, **control_vals)

that appears in our Sage Interact, one can use the following line of code: $f(x)=f$
before we use function f, is in the following example:

```
@interact
def BisectionInteract(
    f = input_box(default = x^2-7, label = 'f(x)' ),
    a = input_box(default = 0 ),
    b = input_box(default = 7 ),
    tolerance = input_box(default= 0.00001) ):
    #include next line to avoid a warning message
    f(x)=f
    x0 = (a+b)/2
    while( f(x0)!=0 and  (b-a) >= tolerance):
        if(f(a)*f(x0)<0) :
            b = x0
        else:              # (f(b)*f(x0)<0)
            a = x0
        x0 = (a+b)/2
    print "approximate solution = ", x0.n()
```

We included this throughout the following examples.

Note: We did not validate the input. If, for example, $f(a)f(b) \geq 0$, and error message would be appropriate. Here is a Sage Interact that would validate the input:

```
@interact
def BisectionInteract(
    f = input_box(default = x^2-7, label = 'f(x)' ),
    a = input_box(default = 0 ),
    b = input_box(default = 7 ),
    tolerance = input_box(default= 0.00001) ):

    f(x)=f
    #validation
    if f(a)*f(b)>=0:
        print "Error: f(a)*f(b) should be negative!"
```

```
else:
    x0 = (a+b)/2
    while( f(x0)!=0 and (b-a) >= tolerance):
        if(f(a)*f(x0)<0):
            b = x0
        else:                # (f(b)*f(x0)<0)
            a = x0
        x0 = (a+b)/2
    print "approximate solution = ", x0.n()
```

And we obtain the output

```
approximate solution = 2.64575147628784
```

Changing the `tolerance` level to `0.0001` we obtain

```
approximate solution = 2.64574813842773
```

```
    f(x)  x^2 - 7

      a  0

      b  7

tollerance  0.0001

    approximate solution =  2.64574813842773
```

If one changes the value of *a* to 3, then we get

```
    f(x)  x^2 - 7

      a  3

      b  7

tollerance  0.0001

    Error: f(a)*f(b) should be negative!
```

To output a list of successive approximation, one simply needs to include a `print` statement inside the `while` loop. Here we use the `round` method in order to avoid printing more than four decimals for the successive approximations.

```
@interact
def BisectionInteract(
    f = input_box(default = x^2-7, label = 'f(x)' ),
    a = input_box(default = 0 ),
```

```
    b = input_box(default = 7 ),
    tolerance = input_box(default= 0.00001) ):

f(x)=f
#validation
if f(a)*f(b)>=0:
    print "Error: f(a)*f(b) should be negative!"

else:
    x0 = (a+b)/2
    while( f(x0)!=0 and  (b-a) >= tolerance):
        print "on [", round(a, 4) , "," ,round(b, 4) ,"]
        \twe have x0 = ",\
                    round( x0, 4), " and f(x0)=", round( f(x0), 7)

        if(f(a)*f(x0)<0):
            b = x0
        else:                   #(f(b)*f(x0)<0)
            a = x0
        x0 = (a+b)/2
    print "approximate solution = ", x0.n()
```

We obtained

```
on [ 0.0 , 7.0 ]      we have x0 = 3.5 and f(x0)= 5.25
on [ 0.0 , 3.5 ]      we have x0 = 1.75 and f(x0)= -3.9375
on [ 1.75 , 3.5 ]      we have x0 = 2.625 and f(x0)= -0.109375
on [ 2.625 , 3.5 ]       we have x0 = 3.0625 and f(x0)= 2.3789063
on [ 2.625 , 3.0625 ]     we have x0 = 2.8438 and f(x0)= 1.0869141
on [ 2.625 , 2.8438 ]     we have x0 = 2.7344 and f(x0)= 0.4768066
on [ 2.625 , 2.7344 ]     we have x0 = 2.6797 and f(x0)= 0.1807251
on [ 2.625 , 2.6797 ]     we have x0 = 2.6523 and f(x0)= 0.0349274
on [ 2.625 , 2.6523 ]     we have x0 = 2.6387 and f(x0)= -0.0374107
on [ 2.6387 , 2.6523 ]     we have x0 = 2.6455 and f(x0)= -0.0012884
on [ 2.6455 , 2.6523 ]     we have x0 = 2.6489 and f(x0)= 0.0168078
on [ 2.6455 , 2.6489 ]     we have x0 = 2.6472 and f(x0)= 0.0077568
on [ 2.6455 , 2.6472 ]     we have x0 = 2.6464 and f(x0)= 0.0032334
on [ 2.6455 , 2.6464 ]     we have x0 = 2.6459 and f(x0)= 0.0009723
on [ 2.6455 , 2.6459 ]     we have x0 = 2.6457 and f(x0)= -0.0001581
on [ 2.6457 , 2.6459 ]     we have x0 = 2.6458 and f(x0)= 0.0004071
on [ 2.6457 , 2.6458 ]     we have x0 = 2.6458 and f(x0)= 0.0001245
on [ 2.6457 , 2.6458 ]     we have x0 = 2.6457 and f(x0)= -1.68e-05
on [ 2.6457 , 2.6458 ]     we have x0 = 2.6458 and f(x0)= 5.39e-05
on [ 2.6457 , 2.6458 ]     we have x0 = 2.6458 and f(x0)= 1.85e-05
approximate solution = 2.64575147628784
```

The next Sage Interact can be used in order to plot the given function as well as each of the successive approximations obtained from the Bisection Method.

```
@interact
def BisectionInteract(
    f = input_box(default = x^2-7, label = 'f(x)' ),
    a = input_box(default = 0 ),
    b = input_box(default = 7 ),
    tolerance = input_box(default= 0.00001) ):
```

```
#save these values for plotting
A = a
B = b

#suppresses a warning message
f(x) = f
#validation
if f(a)*f(b)>=0:
    print "Error: f(a)*f(b) should be negative!"

else:
    x0 = (a+b)/2
    #this list will contain successive approximations
    myList = [(x0,0)]
    while( f(x0)!=0 and (b-a) >= tolerance):
        if(f(a)*f(x0)<0):
            b = x0
        else:                   #(f(b)*f(x0)<0)
            a = x0
        x0 = (a+b)/2
        myList = myList + [(x0,0)]
    #plot the list of points
    p = list_plot(myList, color = 'red', size = 50)
    #plot the function
    q = plot(f(x), x, A, B)
    (p+q).show()
```

We obtain

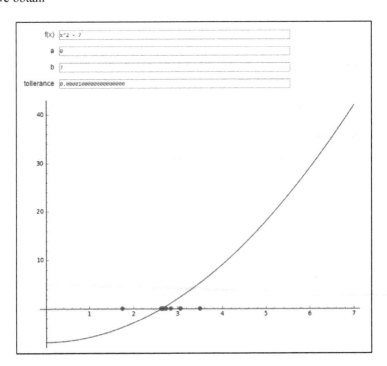

If one chooses the interval from $[0, 7]$ to $[3, 7]$, then

f(x)	x^2 - 7
a	3
b	7
tollerance	0.0000100000000000000

Error: f(a)*f(b) should be negative!

Note: To find an approximate solution for $f(x) = 0$, using the Bisection Algorithm on $[a, b]$, one can use the `bisect` method that is found in the `scipy.optimize` module:

```
import scipy.optimize
@interact
def BisectionInteract(
    f = input_box(default = x^2-7, label = 'f(x)' ),
    a = input_box(default = 0 ),
    b = input_box(default = 7 ),
    tolerance = input_box(default= 0.00001) ):

    #suppresses a warning message
    f(x) = f

    #validation
    if f(a)*f(b)>=0:
        print "Error: f(a)*f(b) should be negative!"

    else:
        print "approximate solution = ", scipy.optimize.bisect(f, a, b)
```

we obtain again

f(x)	x^2 - 7
a	0
b	7
tollerance	0.0000100000000000000

approximate solution = 2.64575131106

4.6.1 Exercises

(1) Modify the interacts given for the Bisection Method so they use a *range slider* to get the user's input for the interval $[a, b]$.

(2) Modify the interacts given for the Bisection Method so they use two *sliders* to get the user's input for the interval $[a, b]$.

4.7 NEWTON–RAPHSON ALGORITHM FOR SOLVING EQUATIONS

Here we again try to solve the equation $f(x) = 0$, either exactly or approximately. In this section, f is assumed to be differentiable.

Newton–Raphson algorithm is described below:

- Pick some initial value to approximate the solution, let's call it x_1.
- Then, using the recursive formula

$$x_n = x_{n-1} - \frac{f(x_{n-1})}{f'(x_{n-1})}, \quad n = 2, 3, 4, \cdots$$

we compute $x_2, x_3, x_4, \cdots$.

- We stop when
  - either x_n is good enough (or it is an exact solution)
  - or we tried too many steps (remember that the sequence may or may not converge!)
  - or the value of the $f'(x_{n-1})$ is too close to 0 for some value of n.

See [2] or [7] for more details on this topic.

We give below our Sage Interact that implements the Newton–Raphson Algorithm.

```
@interact
def NewtonRaphsonInteract(
    f = input_box(default = x^2-7, label = 'f(x)' ),
    x1 = input_box(default = 3 ),
    max_steps = input_box(default = 100 ),
    tolerance = input_box(default= 0.00001) ):

    #suppresses a warning message
    f(x) = f

    #we start with the initial value x1
    xn = x1

    #we need the derivative of f with respect to x
    fprime(x) = derivative(f(x),x)
```

```
steps = 1

#keep computing successive approximations as long as
# we did not reach max_steps
# we do not have a reasonable solution and
# the derivative is not too close to zero
while(   steps<=max_steps and          \
         abs(f(xn))>tolerance and \
         abs(fprime(xn))>tolerance  ) :
    xn -= f(xn)/fprime(xn)
    steps +=  1
if(abs(fprime(xn))<=tolerance ) :
    print "error!!! derivative value is too close to zero", xn.n()
elif (steps>=max_steps) :
    print "I gave up ... too many steps!"
else :
    print "approximate solution = ", xn.n()
```

We obtain the solution

```
approximate solution = 2.64575131233596
```

and the Sage Interact looks like

If the user chooses a smaller tolerance level, we obtain

which is a more accurate answer, closer to the output of the following value:

```
sqrt(7).n()
```

which yields

```
2.64575131106459
```

Choosing x_1 to be 0, we obtain

```
     f(x)  x^2 - 7

       x1  0

 max_steps  100

 tollerance  0.0000100000000000000

   error!!! derivative value is too close to zero 0.000000000000000
```

The next Sage Interact can be used in order to plot the given function as well as each of the approximations considered.

```
@interact
def NewtonRaphsonInteract2(
      f = input_box(default = x^2-7, label = 'f(x)' ),
      x1 = input_box(default = 3 ),
      max_steps = input_box(default = 100 ),
      tolerance = input_box(default= 0.00001) ):

      #suppresses a warning message
      f(x) = f

      #we start with x1
      xn = x1

      #this will collect the list of all successive approximations
      myList = [(xn,0)]

      #we need the derivative of f with respect to x
      fprime(x) = derivative(f(x),x)

      steps = 1

      #keep computing successive approximations as long as
      # we did not reach max_steps
      # we do not have a reasonable solution and
```

```
# the derivative is not too close to zero
while(  steps<=max_steps and          \
        abs(f(xn))>tolerance and \
        abs(fprime(xn))>tolerance  ):
    xn -= f(xn)/fprime(xn)
    myList += [(xn,0)]
    steps +=  1

#output the results
if(abs(fprime(xn))<= tolerance ):
    print "error, derivative value close to zero", xn.n()
elif (steps>=max_steps):
    print "I gave up ... too many steps!"
else:
    print "approximate solution = ", xn.n()

#make the plot either way - so it can be analyzed
#    -  plot the list of points
p = list_plot(myList, color = 'red', size = 50)
#    -  plot the function
a = min([myList[i][0] for i in [0,1,..,len(myList)-1]])
b = max([myList[i][0] for i in [0,1,..,len(myList)-1]])
q = plot(f(x), x, a-5, b+5)
(p+q).show()
```

We obtain

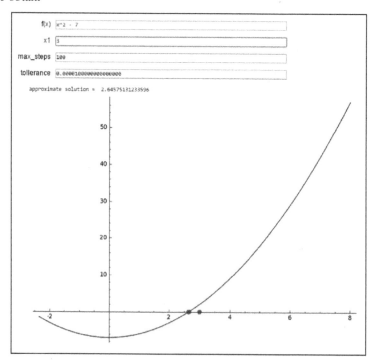

Starting with a worse approximation $x_1 = 300$, and a lower tolerance level, we obtain

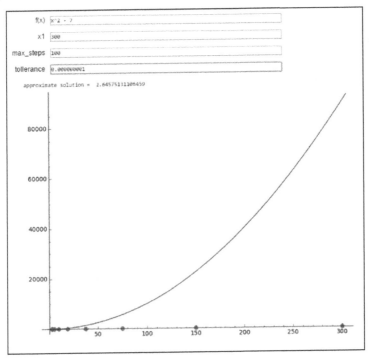

The next Sage Interact will use a slider to select the number of iterations for the Newton–Raphson algorithm, and then plot these points:

```
@interact
def NewtonRaphsonInteract3(
    f = input_box(default = x^2-7, label = 'f(x)' ),
    x1 = input_box(default = 30 ),
    iterations = slider(vmin=1, vmax=10,
                        step_size=1,
                        default=1),
    tolerance = input_box(default= 0.00001) ):

    #suppresses a warning message
    f(x) = f

    #we start with x1
    xn = x1

    #this will collect the list of all successive approximations
    myList = [(xn,0)]

    #we need the derivative of f with respect to x
    fprime(x) = derivative(f(x),x)
```

```
steps = 1

#keep computing successive approximations as long as
# we did not reach max_steps
# we do not have a reasonable solution and
# the derivative is not too close to zero
while(   steps<=iterations and            \
         abs(f(xn))>tolerance and \
         abs(fprime(xn))>tolerance   ):
    xn -= f(xn)/fprime(xn)
    myList += [(xn,0)]
    steps += 1

#output the results
if(abs(fprime(xn))<= tolerance ):
    print "error, derivative value close to zero", xn.n()
else:
    print "approximate solution = ", xn.n()

#make the plot either way - so it can be analyzed
#  -  plot the list of points
p = list_plot(myList, color = 'red', size = 50)
#  -  plot the function
a = min([myList[i][0] for i in [0,1,..,len(myList)-1]])
b = max([myList[i][0] for i in [0,1,..,len(myList)-1]])
q = plot(f(x), x, a-5, b+5)
(p+q).show()
```

We obtain

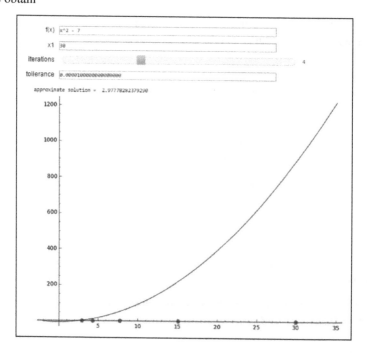

To include a label for each iteration, one can use the following variation of the previous Sage Interact:

```
@interact
def NewtonRaphsonInteract4(
    f = input_box(default = x^2-7, label = 'f(x)' ),
    x1 = input_box(default = 30 ),
    iterations = slider(vmin=1, vmax=10,
                        step_size=1,
                        default=1),
    tolerance = input_box(default= 0.00001) ):

    #suppresses a warning message
    f(x) = f

    #we start with x1
    xn = x1

    #this will collect the list of all successive approximations
    myList = [(xn,0)]

    #we need the derivative of f with respect to x
    fprime(x) = derivative(f(x),x)

    steps = 1

    #keep computing successive approximations as long as
    # we did not reach max_steps
    # we do not have a reasonable solution and
    # the derivative is not too close to zero
    while( steps<=iterations and        \
           abs(f(xn))>tolerance and \
           abs(fprime(xn))>tolerance ):
        xn -= f(xn)/fprime(xn)
        myList += [(xn,0)]
        steps += 1

    #output the results
    if(abs(fprime(xn))<= tolerance ):
        print "error, derivative value close to zero", xn.n()
    else:
        print "approximate solution = ", xn.n()

    #make the plot either way - so it can be analyzed
    #   -  plot the list of points
    p = list_plot(myList, color = 'red', size = 50)

    #   -  to put the text, we need to find how big is the graph
    y1 = min([f(myList[i][0]) for i in [0,1,..,len(myList)-1]])
    y2 = max([f(myList[i][0]) for i in [0,1,..,len(myList)-1]])
    posT = y1-0.10*(y2-y1)
    t = text(1, (myList[0][0], posT), fontsize=20,       \
             color = 'black')
```

```
for i in range(1, len(myList)):
    t = t+ text(i+1, (myList[i][0],posT), fontsize=20, \
                color = 'black')

#  -  plot the function
a = min([myList[i][0] for i in [0,1,..,len(myList)-1]])
b = max([myList[i][0] for i in [0,1,..,len(myList)-1]])
q = plot(f(x), x, a-5, b+5)
(p+q+t).show()
```

and obtain a Sage Interact similar to

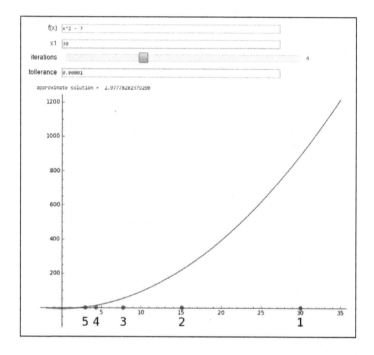

For other values, one obtains the following:

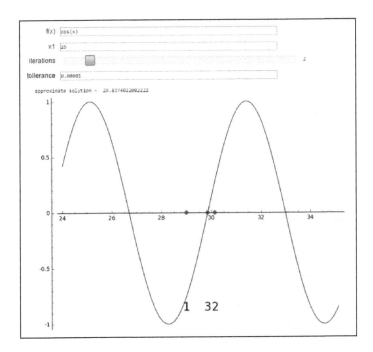

To include the tangent line in the graph, one can use

```
@interact
def NewtonRaphsonInteract5(
    f = input_box(default = x^2-7, label = 'f(x)' ),
    x1 = input_box(default = 30 ),
    iterations = slider(vmin=1, vmax=10,
                        step_size=1,
                        default=1),
    tolerance = input_box(default= 0.00001) ):
```

```
#suppresses a warning message
f(x) = f

#we start with x1
xn = x1

#this will collect the list of all successive approximations
myList = [(xn,0)]

#we need the derivative of f with respect to x
fprime(x) = derivative(f(x),x)

steps = 1

#keep computing successive approximations as long as
# we did not reach max_steps
# we do not have a reasonable solution and
# the derivative is not too close to zero
while(  steps<=iterations and         \
        abs(f(xn))>tolerance and \
        abs(fprime(xn))>tolerance  ):
    xn -= f(xn)/fprime(xn)
    myList += [(xn,0)]
    steps += 1

#output the results
if(abs(fprime(xn))<= tolerance ):
    print "error, derivative value close to zero", xn.n()
else:
    print "approximate solution = ", xn.n()

#make the plot either way - so it can be analyzed
# -  plot the list of points
p = list_plot(myList, color = 'red', size = 50)
# -  to put the text, we need to find how big is the graph
y1 = min([f(myList[i][0]) for i in [0,1,..,len(myList)-1]])
y2 = max([f(myList[i][0]) for i in [0,1,..,len(myList)-1]])
posT = y1-0.10*(y2-y1)
t = text(1, (myList[0][0], posT), fontsize=20, color = 'black')

for i in range(1, len(myList)):
    t = t+ text(i+1, (myList[i][0],posT), fontsize=20,
    color = 'black')
    X1 = myList[i-1][0]
    Y1 = f(X1)
    X2 = myList[i][0]
    Y2 = f(X2)
    #tangent line
    t = t + line([(X1, Y1),(X2,0)], color = 'green')
    #vertical line
    t = t + line([(X2, 0),(X2,Y2)], color = 'green')
# -  plot the function
```

```
a = min([myList[i][0] for i in [0,1,..,len(myList)-1]])
b = max([myList[i][0] for i in [0,1,..,len(myList)-1]])
q = plot(f(x), x, a-5, b+5)
(p+q+t).show()
```

and obtain

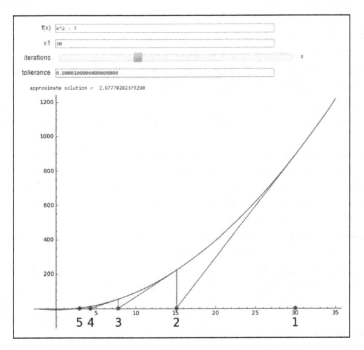

When a function is not differentiable, or when its derivative is not easily computable, one can use an algorithm called the **Secant Algorithm**. In this algorithm, we do not need to require the function f to be differentiable. The implementation of it using Sage Interacts is left as an exercise. Here is the algorithm:

- Pick two initial values to approximate the solution, let's call them x_1 and x_2.
- Then, using the recursive formula

$$x_n = x_{n-1} - f(x_{n-1}) \cdot \frac{x_{n-1} - x_{n-2}}{f(x_{n-1}) - f(x_{n-2})}, \quad n = 3, 4, 5, \ldots .$$

We compute $x_3, x_4, x_5, \cdots$.

- We stop when either x_n is good enough (or exact solution), or we tried too many steps (remember that the sequence may or may not converge!).

See [2; 7] for more details on this topic.

4.7.1 Exercises

(1) Modify the interacts given for the Newton–Raphson Method so they use a *slider* to get the user's input for the interval initial point x_1.

(2) Modify the interacts given for the Newton–Raphson Method so they allow you to use two initial points and will display two parallel results (one for each of the two initial points given by the user).

(3) Create a Sage Interact that will output each successive approximation of the Newton–Raphson Method, for a function, an initial value, a tolerance level, and a maximum number of steps, all given by the user.

(4) Modify the Sage Interact obtained in the aforementioned exercise so that it uses a slider for the maximum number of iterations the algorithm can run.

(5) Find an implementation of the Newton–Raphson Method in either SciPy or NumPy libraries. Then, use them to simplify the Sage Interacts given in this section.

(6) Implement the Secant Algorithm using Sage Interacts. The Interact should allow the user to input a function, two distinct initial points, a maximum number of steps allowed for the algorithm to run, and a tolerance level. Default values should be $\sin(x)$, 3, 4, 1000, and 0.00001, respectively. It should output the solution or an error message if needed.

(7) Modify the code obtained at the previous problem so that it outputs each of the successive approximations of the solution.

(8) Implement the Secant Algorithm using Sage Interacts. The Interact should allow the user to input a function, two distinct initial points, a maximum number of steps allowed for the algorithm to run, and a tolerance level. Default values should be $\sin(x)$, 3, 4, 1000, and 0.00001, respectively. It should plot the function and the list of successive approximations.

(9) Modify the code obtained at the previous problem so that it plots not only the function and the successive approximations of the solution, but it also plots the secant lines.

(10) Find an implementation of the Secant Algorithm in either SciPy or NumPy libraries. Then, use them to create Sage Interacts similar to the ones given in this section.

(11) Using a Numerical Analysis book as a guide, implement the Fixed-Point Algorithm using Sage Interacts. The Interact should allow the user to input a function, an initial point, a maximum number of steps allowed for the algorithm to run, and a tolerance value. Default values should be $\sin(x)$, 3, 1000, and 0.00001, respectively. It should output the solution or an error message if needed.

(12) Modify the code obtained at the previous problem so that it outputs each of the successive values.

(13) Using a Numerical Analysis book as a guide, implement the Fixed-Point Algorithm using Sage Interacts. The Interact should allow the user to input a function, an initial point, a maximum number of steps allowed for the algorithm to run, and a tolerance level. Default values should be $\sin(x)$, 3, 1000, and

0.00001, respectively. It should plot the function and the list of successive approximations.

(14) Modify the code obtained at the previous problem so that it plots not only the function and the successive approximations of the fixed point, but it also plots the vertical and horizontal lines that demonstrate the successive steps.

(15) Find an implementation of the Fixed-Point Algorithm in either SciPy or NumPy libraries. Then, use them to create Sage Interacts similar to the ones given in this section.

4.8 POLYNOMIAL INTERPOLATION

Here we only give Sage implementation of the Polynomial Interpolation using **Lagrange Form**. For other forms (such as Newton form, or Divided Differences), and for more details about Lagrange Form, see [2] or [7].

Given a set of $n + 1$ distinct data points

x_0	x_1	x_2	$\cdots$	x_n
y_0	y_1	y_0	$\cdots$	y_n

one can find the unique polynomial of degree at most n that passes through these $n + 1$ distinct points as follows:

$$p(x) = y_0 L_0 + y_1 L_1 + \cdots + y_n L_n,$$

where

$$L_k = \frac{(x - x_0) \cdots (x - x_{k-1})(x - x_{k+1}) \cdots (x - x_n)}{(x_k - x_0) \cdots (x_k - x_{k-1})(x_k - x_{k+1}) \cdots (x_k - x_n)}$$

for $k = 0, 1, 2, \ldots, n$.

We give below, a Sage Interact that allows the user to input a set of points, and it computes the polynomial of least degree that interpolates them:

```
def Lagrange_Basis(i, points):
    var('x')
    n = len(points)-1

    Li=1
    for j in [0,1,..,n] :
        if(i!=j):
            Li *= (x-points[j][0])/(points[i][0]-points[j][0])
    return Li


def Lagrange_Polynomial(points):
    var('x')
    n = len(points)-1

    p=0*x
    for i in [0,1,..,n]:
```

```
      p +=  points[i][1]*Lagrange_Basis(i,points)
   return p.full_simplify()

@interact
def LagrangeInterpolationInteract(
   points = input_box(default = [(-3,-15),(-1,-5),(0,1),(2,10),
   (3,15)]) ):
   print "p(x)=", Lagrange_Polynomial(points)
```

We obtain a Sage Interact similar to the one shown below:

points [(-3, -15), (-1, -5), (0, 1), (2, 10), (3, 15)]

p(x)= 1/18*x^4 - 1/18*x^3 - 11/18*x^2 + 11/2*x + 1

The following Sage Interact will plot the given points and the polynomial of interpolation:

```
@interact
def LagrangeInterpolationInteract2(
   points = input_box(default = [(-3,-15),(-1,-5),(0,1),(2,10),
   (3,15)]) ):

   n = len(points)-1
   a = min([points[i][0] for i in [0,1,..,n]])
   b = max([points[i][0] for i in [0,1,..,n]])
   p = plot(Lagrange_Polynomial(points) , x, a-1, b + 1)
   p = p + list_plot( points, color = 'red', size = 70)
   show (p)
```

We obtain

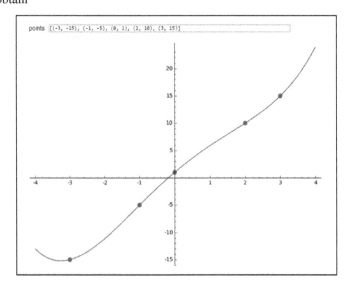

The following variation of the previous Sage program uses a slider that selects how many points from the given list to use for the Polynomial Interpolation and plots this polynomial. Then it plots all the given points. Since we could not find an easy way to automate the range of the slider to fit the exact number of points given in the list of points, we hard-coded it as a maximum of 10. At the end of this section, we show how one can use **nested interacts** to perform this.

```
@interact
def LagrangeInterpolationInteract3(
    points = input_box(default =  [(-3,-15), (-1,-5), (0, 1), (2,10),
    (3,15)] ) ,
    k = slider(vmin=1,vmax=10 , step_size=1, label="#of data points
    to use: ") ):
  n = len(points)-1
  if k<=len(points):
      a = min([points[i][0] for i in [0,1,..,n]])
      b = max([points[i][0] for i in [0,1,..,n]])
      sublist =  [points[i] for i in [0,1,..,k-1]]
      p = plot(Lagrange_Polynomial(sublist) , x, a-4, b + 4)
      p = p + list_plot( points, color = 'red', size = 70)
      show (p)
  else:
      print "selected value of k:", k
      print "#data should not be larger than the actual # of points"
```

We obtained the following:

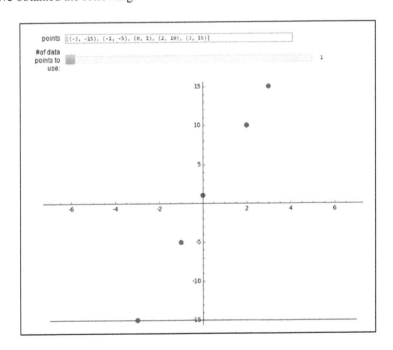

Changing the value of the slider, we get

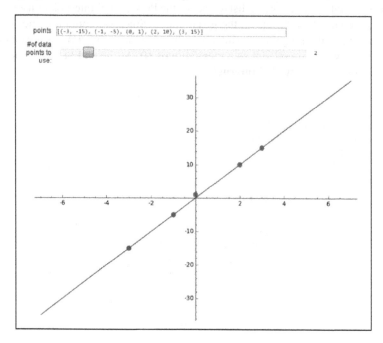

and

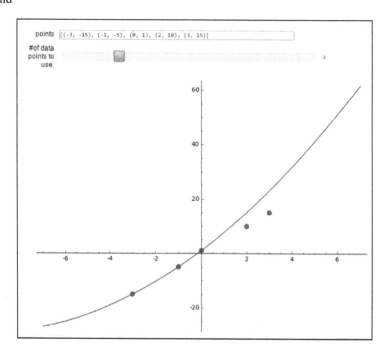

If we happen to use the slider and choose a value larger than the number of points, then we should get an error message:

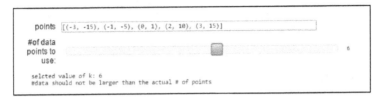

Note: One can also use a `for` loop to provide the list of points as in the following example:
```
[ (x, sin(x) ) for x in [1,2,..,10]]
```
and obtain

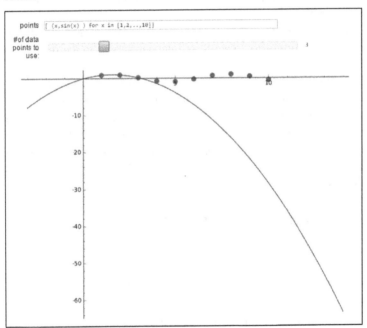

One can modify the previous Sage Interact and use nested interacts:

```
@interact
def LagrangeInterpolationInteract3(
    points = input_box(default =  [(-3,-15), (-1,-5), (0, 1), (2,10),
    (3,15)] )):

    n = len(points)-1

    @interact
    def NestedInteract(
        k = slider(vmin=1,vmax=n+1,
            step_size=1, label="#of data points to use: ") ):
```

```
a = min([points[i][0] for i in [0,1,..,n]])
b = max([points[i][0] for i in [0,1,..,n]])
sublist = [points[i] for i in [0,1,..,k-1]]
p = plot(Lagrange_Polynomial(sublist) , x, a-4, b + 4)
p += list_plot( points, color = 'red', size = 70)
show (p)
```

4.8.1 Exercises

(1) Using a Numerical Analysis book as a guide, implement a Sage Interact for computing the Interpolating Polynomial using the Newton Form of Polynomial Interpolation.

(2) Using a Numerical Analysis book as a guide, implement a Sage Interact that graphs the Interpolating Polynomial obtained using the Newton Form of Polynomial Interpolation.

(3) Find an implementation of the Newton Form algorithm in either SciPy or NumPy libraries. Then, use it to create Sage Interacts similar to the ones given in this section.

(4) Using a Numerical Analysis book as a guide, implement a Sage Interact for computing the Interpolating Polynomial using Divided Differences.

(5) Using a Numerical Analysis book as a guide, implement a Sage Interact that graphs the Interpolating Polynomial obtained using Divided Differences.

(6) Find an implementation of the Divided Differences algorithm in either SciPy or NumPy libraries. Then, use it to create Sage Interacts similar to the ones given in this section.

4.9 LINEAR SPLINE INTERPOLATION

Given a set of $n + 1$ distinct data points

x_0	x_1	x_2	$\cdots$	x_n
y_0	y_1	y_0	$\cdots$	y_n

one can find a continuous piecewise-defined function, whose "pieces" are linear functions:

$$L(x) = \begin{cases} m_1 x + b_1 & \text{for} \quad x \in [x_0, x_1] \\ m_2 x + b_2 & \text{for} \quad x \in [x_1, x_2] \\ \quad \vdots \\ m_n x + b_n & \text{for} \quad x \in [x_{n-1}, x_n] \end{cases}.$$

See [2] or [7] for more details about this topic.

We give below, a Sage Interact that allows the user to input a set of points, and it draws the corresponding linear spline:

```
#compute and draw the linear spline
def Linear_Spline(points):
    var('x')
```

```
#first we plot the given points
p = list_plot( points, color = 'red', size = 50)

#compute the number of subintervals
n = len(points)-1

#then compute and plot each branch of the linear spline
for k in [1,..,n]:
    #we find the equation y=mx+b for the kth subinterval
    m = (points[k][1]-points[k-1][1])/ \
        (points[k][0]-points[k-1][0])
    b = points[k][1]-m*points[k][0]

    #then we plot that line on the interval [xk-1, xk]
    p += plot(m*x+b, x,points[k-1][0],points[k][0],
              color = 'green',
              title = 'linear spline interpolation')

#display the linear spline and the given points
p.show()

@interact
def LinearSplineInteract(
    points = input_box(default = [(-3,-10),(-1,-5),(1,1),(2,10),
    (4,15)]) ):
    Linear_Spline(points)
```

We obtain

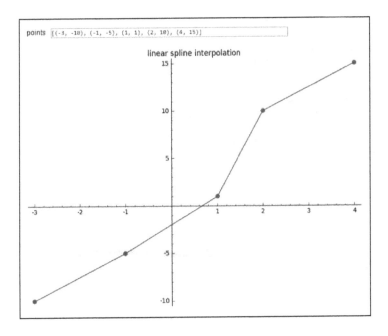

Using the following set `[(x,sin(x)) for x in [0, pi/4,..,2*pi]]`
we obtain

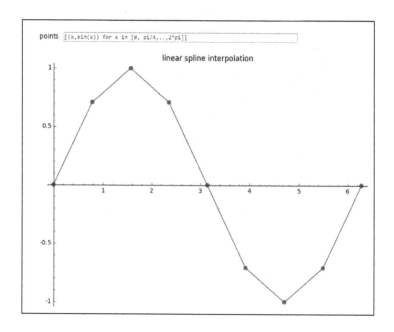

We give, next, an implementation of the previous code using SciPy libraries.

```
from scipy.interpolate import interp1d

def Linear_Spline(points):
    #compute the size of the given array
    n = len(points)

    #get the x values of points
    xvals = [points[i][0] for i in [0,1,..,n-1]]

    #get the y values of points
    yvals = [points[i][1] for i in [0,1,..,n-1]]

    #find the linear interpolation
    f=interp1d(xvals, yvals)

    #find the interval [a,b]
    a = min([points[i][0] for i in [0,1,..,n-1]])
    b = max([points[i][0] for i in [0,1,..,n-1]])
```

```
#plot the linear interpolation
p = plot(f, x, a, b, color = 'green',
     title = 'linear spline interpolation')

#plot the points
p += list_plot( points, color = 'red', size = 50)

p.show()
```

```
@interact
def LinearSplineInteract(
    points = input_box(default = [(-3,-10),(-1,-5),(1,1),(2,10),
    (4,15)]) ):
    Linear_Spline(points)
```

The output is the same as the ones given above, therefore we did include them again.

Note: Replacing the line

```
f=interp1d(xvals, yvals)
```

with

```
f=interp1d(xvals, yvals, kind = 'cubic')
```

one obtains a cubic spline as shown below:

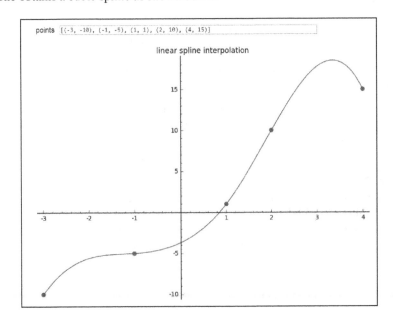

4.9.1 Exercises

(1) Create a Sage Interact that allows the user to enter a function f, an interval $[a, b]$, and a number n. Then, the interact will graph the function f and the spline function that interpolates the function f at n equally spaced points in the interval $[a, b]$.

(2) Create a Sage Interact that allows the user to enter a function f and a set of points. Then, the interact will graph the function f and the spline function that interpolates the function f at the given set of points. Hint: From the given set of points, you will need to compute the interval $[a, b]$ that will be used for graphing the given function and the computed spline function.

(3) Using a Numerical Analysis book as a guide, implement a Sage Interact for the Quadratic Spline Interpolation.

(4) Modify the Sage Interact obtained in the previous exercise so that the user can also select the value of $QS'(x_0)$ using a slider with numbers between $[-3, 3]$.

(5) Create a Sage Interact that allows the user to input a set of points and select whether to plot the linear spline, cubic spline, or both. Regardless of the choice, the interact should plot the data points given.

(6) Extend the Sage Interact created in the previous exercise so that it allows the user to specify what colors to use for each one of the splines.

4.10 CUBIC SPLINE INTERPOLATION

Given a set of $n + 1$ distinct data points

$$\begin{array}{c|c|c|c|c}
x_0 & x_1 & x_2 & \cdots & x_n \\
\hline
y_0 & y_1 & y_0 & \cdots & y_n
\end{array},$$

one can find a continuous piecewise-defined function, whose pieces are cubic polynomial functions:

$$C(x) = \begin{cases}
a_1 x^2 + b_1 x + c_1 & \text{for } x \in [x_0, x_1] \\
a_2 x^2 + b_2 x + c_2 & \text{for } x \in [x_1, x_2] \\
\quad \vdots \\
a_n x^2 + b_n x + c_n & \text{for } x \in [x_{n-1}, x_n]
\end{cases}.$$

See [2] or [7] for more details on this topic.

Note: To compute the spline function for a given set of data points, one can use the Sage library function: `spline`.

For simplicity, in this section, we continue with the SciPy library function `interp1d` that was introduced in the previous section.

```
from scipy.interpolate import interp1d

#compute and draw the cubic spline
def Cubic_Spline(points):
    #needed if you want to use x in the input
    # like  [(x, sin(x)) for x in [0,1,..,4]]
    var('x')
```

```
#compute the size of the given array
n = len(points)

#get the x values of points
xvals = [points[i][0] for i in [0,1,..,n-1]]

#get the y values of points
yvals = [points[i][1] for i in [0,1,..,n-1]]

#find the cubic spline interpolation
f=interp1d(xvals, yvals, kind = 'cubic')

#find the interval [a,b]
a = min([points[i][0] for i in [0,1,..,n-1]])
b = max([points[i][0] for i in [0,1,..,n-1]])

#plot the linear interpolation
p = plot(f, x, a, b, color = 'green',
    title = 'linear spline interpolation')

#plot the points
p += list_plot( points, color = 'red', size = 50)

p.show()
```

```
@interact
def CubicSplineInteract(
    points = input_box(default = [(-3,-10),(-1,-5),(1,1),(2,10),
    (4,15)]) ):
    Cubic_Spline(points)
```

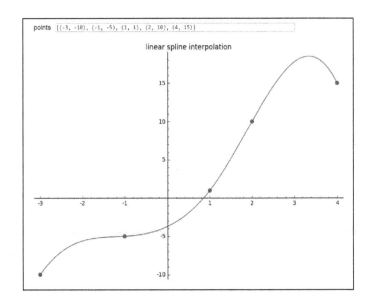

Note: Using the Sage method `spline`, one would obtain the following cubic spline:

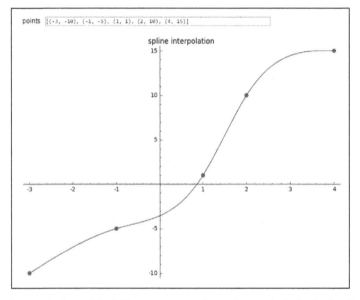

To understand where this difference comes from, the reader is invited to find more about *natural cubic splines*, *B-splines*, and *complete cubic splines* in a specialized Numerical Analysis book such as [2] or [7].

Using the Sage library function spline, one could use the following set: `[(x,sin(x)) for x in [0, pi/4,..,2*pi]]` in order to obtain

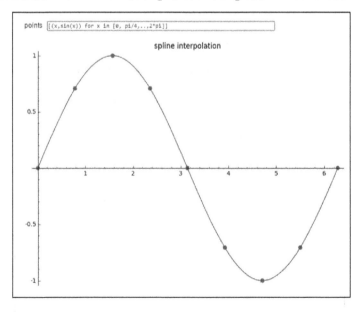

The way our Sage Interact is given above, such an input will result in the following error:

```
TypeError: Cannot cast array data from dtype('O') to
dtype('float64') according to the rule 'safe'
```

The reason for this is that the function `interp1d` does not recognize pi. Hence, to address this, one needs to include the following line at the beginning of the code given above:

```
import scipy
```

and then one can use the following input:

```
[(x,sin(x)) for x in [0, scipy.pi/4,..,2*scipy.pi]]
```

and obtain the same output as the one given in the image above, namely

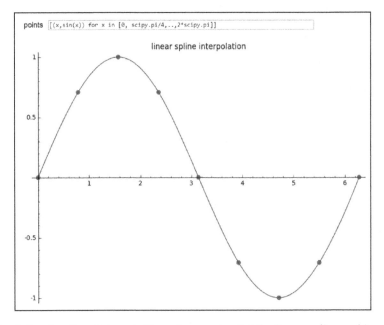

The following Sage Interact allows the user to plot the linear spline, cubic spline, or both on the same graph.

```
import scipy
from scipy.interpolate import interp1d

#compute and draw the cubic spline
def Linear_Cubic_Spline(points, what_to_plot):
    #needed if you want to use x in the input
```

```
    # like  [(x, sin(x)) for x in [0,1,..,4]]
    var('x')

    #compute the size of the given array
    n = len(points)

    #get the x values of points
    xvals = [points[i][0] for i in [0,1,..,n-1]]

    #get the y values of points
    yvals = [points[i][1] for i in [0,1,..,n-1]]

    #plot the points
    p = list_plot( points, color = 'red', size = 50)

    #find the interval [a,b]
    a = min([points[i][0] for i in [0,1,..,n-1]])
    b = max([points[i][0] for i in [0,1,..,n-1]])

    if(what_to_plot == "Linear"):
       f1=interp1d(xvals, yvals, kind = 'linear')
       p += plot(f1, x, a, b, color = 'green',
         legend_label = 'linear spline interpolation')

    elif(what_to_plot == "Cubic"):
       f3=interp1d(xvals, yvals, kind = 'cubic')
       p += plot(f3, x, a, b, color = 'blue',
         legend_label = 'cubic spline interpolation')

    elif(what_to_plot == "Both"):
       f1=interp1d(xvals, yvals, kind = 'linear')
       p += plot(f1, x, a, b, color = 'green',
         legend_label = 'linear spline interpolation')

       f3=interp1d(xvals, yvals, kind = 'cubic')
       p += plot(f3, x, a, b, color = 'blue',
         legend_label = 'cubic spline interpolation')
    else:
        print "ERROR!"

    #return the plot object
    return p

@interact
def SplinesInteract(
    points = input_box(default = [(-3,-10),(-1,-5),(1,1),(2,10),
    (4,15)]) ,
    plots = selector(values = ["Linear", "Cubic", "Both"],
                    label = "Select the plots",default = "Both" )):
   Linear_Cubic_Spline(points, plots).show()
```

We obtain

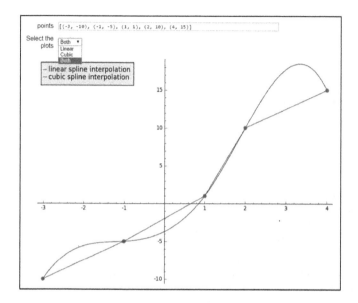

The following Sage Interact allows the user to use checkboxes to select whether to plot the polynomial interpolant, linear spline, and/or cubic spline for a given set of data points:

```
import scipy
from scipy.interpolate import interp1d

def Lagrange_Basis(i, points):

    given on page 179


def Lagrange_Polynomial(points):

    given on page 179


def DrawInterpolations(points, plotPoly, plotLinear, plotCubic):
    #needed if you want to use x in the input
    # like  [(x, sin(x)) for x in [0,1,..,4]]
    var('x')
```

```
    #compute the size of the given array
    n = len(points)

    #get the x values of points
    xvals = [points[i][0] for i in [0,1,..,n-1]]

    #get the y values of points
    yvals = [points[i][1] for i in [0,1,..,n-1]]

    #plot the points
    p = list_plot( points, color = 'red', size = 50)

    #find the interval [a,b]
    a = min([points[i][0] for i in [0,1,..,n-1]])
    b = max([points[i][0] for i in [0,1,..,n-1]])

    if(plotPoly == True):
        fp=Lagrange_Polynomial(points)
        p += plot(fp, x, a, b, color = 'purple',
            legend_label = 'polynomial interpolation')

    if(plotLinear == True):
        f1=interp1d(xvals, yvals, kind = 'linear')
        p += plot(f1, x, a, b, color = 'green',
            legend_label = 'linear spline interpolation')

    if (plotCubic == True):
        f3=interp1d(xvals, yvals, kind = 'cubic')
        p += plot(f3, x, a, b, color = 'blue',
            legend_label = 'cubic spline interpolation')

    #return the plot object
    return p

@interact
def InterpolationsInteract(
    points = input_box(default = [(-3,-10),(-1,-5),(1,1),(2,10),
    (4,15)]) ,
    plot_poly = ("Polynomial Interpolation", true),
    plot_linear = ("Linear Spline", true),
    plot_cubic = ("Cubic Spline", true)      ):

    DrawInterpolations(points, plot_poly, plot_linear, plot_cubic)
    .show()
```

We obtain

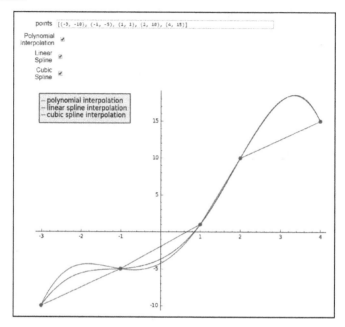

and, for `[(x, sin(x)) for x in [0, scipy.pi/2,..,4*scipy.pi]]`:

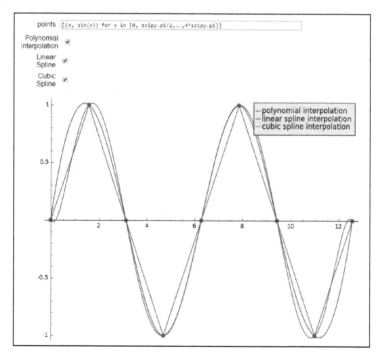

To see advantages of cubic spline interpolation versus Polynomial Interpolation, analyze the results for following set of data points [(-2, 1), (-1, 1), (1, 1.001), (3, 0.999), (7, 1), (10, 1)]:

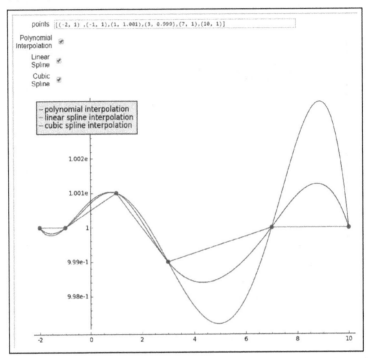

4.10.1   Exercises

(1) Modify the Sage Interacts given in this section so that it uses the Sage `spline` function.

(2) Modify the last Sage Interact given in this section so that it also plots the quadratic polynomial. Hint: Use `kind = 'quadratic'` in the `interp1d` function.

4.11   SAGE FOR SOLVING DIFFERENTIAL EQUATIONS

One can use Sage methods to solve **some** ordinary differential equations (ODEs).
   For example, to solve the following IVP

$$x' + x = 2,$$

$$x(0) = 10,$$

where $x = x(t)$, one can use the following Sage code:

```
var('t')
x = function('x', t)
de = diff(x,t) + x - 2
desolve(de==0, x, ics = [0,10])
```

and obtain the solution

```
2*(e^t + 4)*e^(-t)
```

Removing `ics = [0,10]` from the last line of Sage code, one can solve the more general ODE problem

$$x' + x = 2$$

and obtain

```
(_C + e^t)*e^(-t)
```

where `_C` is a placeholder for any constant.

We have given next a Sage Interact that will allow the user to specify an ODE to solve and an optional boundary condition:

```
var('t')
x = function('x', t)

@interact
def ODEInteract(
    choice = selector(values = ['yes','no'],
                      label = 'initial condition?',
                      default = 'yes')):
    if (choice == 'yes'):
        @interact
        def Interact1( de = input_box(default = "diff(x,t) + x - 2" ,
                                      label = "ODE = "),
                       t0 = input_box(default = 0,
                                      label = "t0="),
                       xt0 = input_box(default = 10,
                                       label = "x(t0)=")):
            desolve(de==0, x, ics = [t0, xt0]).show()
    else:
        @interact
        def Interact2( de = input_box(default = "diff(x,t) + x - 2" ,
                       label = "ODE = ")):
            desolve(de==0, x).show()
```

We obtained

initial
condition? no ▼

    ODE = `diff(x(t),t) + x(t) - 2` ★

$$\left(C + 2\,e^{t}\right)e^{(-t)}$$

and

initial
condition? yes ▼

    ODE = `diff(x(t),t) + x(t) - 2` ★
    t0= `0`
    x(t0)= `10`

$$2\left(e^{t} + 4\right)e^{(-t)}$$

Changing some values, we get

initial
condition? yes ▼

    ODE = `diff(x(t),t) + x(t) - 10` ★
    t0= `20`
    x(t0)= `100`

$$10\left(9\,e^{20} + e^{t}\right)e^{(-t)}$$

Note: In the aforementioned examples, one could also have used the following: `derivative(x,t) + x - 10`.

The next code is a simplified version of the previous Sage Interact:

```
var('t')
x = function('x', t)
```

```
@interact
def ODEInteract2(
            de = input_box(default = "diff(x,t,t) + x - 2" ,
                                  label = "ODE = ")):
            desolve(de==0, x).show()
```

One can use it to solve differential equations such as

$$x'' + x = 2$$

and get

ODE =	diff(x,t,t) + x - 2
	$K_2 \cos(t) + K_1 \sin(t) + 2$

At the moment of writing this book, Sage capabilities for solving ODE and partial differential equation (PDE) are quite limited. Try, for example, to solve the following differential equation:

$$(x'')^2 = 4.$$

For such reasons, numerical approximations can come very handy. The following two sections are focusing on these topics.

4.12 NUMERICAL METHODS FOR ORDINARY DIFFERENTIAL EQUATIONS

In this section, we implement Euler Method for Differential Equations using Sage Interacts. For more details about this topic, please refer to [2] or [7].

One can use numerical methods in order to approximate the solution of the following ODE (IVPs of first-order derivatives):

$$\begin{cases} y' = f(x, y) & \text{for } x \in [a, b] \\ \quad y(a) = y_0 \end{cases}.$$ (4.1)

The solution is given in form of a table of values, for the $n + 1$ nodes $x_0 = a, x_1 = a + h, x_2 = a + 2h, \cdots, x_n = b$, where

$$h = \frac{b - a}{n}.$$

Using interpolation techniques, one can then approximate the solution y for other values of x than the nodes.

Here comes the algorithm called **Euler Method**:

$$y(x_0) = y_0,$$

$$y(x_i) = y(x_{i-1}) + hf(x_{i-1}, y(x_{i-1})) \quad \text{for } i = 1, 2, \cdots, n.$$

Next we give a Sage Interact that creates and plots such a table:

```
var('x,y')
@interact
def EulerDEInteract(
    f = input_box(default= cos(x)+sin(y)+x*y),
    n = slider(vmin=0, vmax=100, step_size=1,
                default=3,
                label="Select the order n: "),
    a = input_box(default= 0),
    b = input_box(default = 2),
    y0= input_box(default = 8)):

    #this it to avoid a warning message
    f(x,y)=f

    #compute the step size
    h = (b-a)/n

    #create the table of approximations
    table = [(a, y0)]

    prevx = a
    prevy = y0

    for i in [1,2,..,n]:
        #compute the new values
        newx = prevx+h
        newy = prevy + h*f(prevx, prevy)

        #add them the to the table
        table += [(newx, newy)]

        #"move" to the new values
        prevx = newx.n()
        prevy = newy.n()

    #plot the obtained table
    list_plot(table,size = 50).show()
```

We obtain

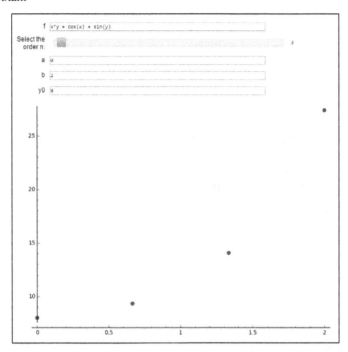

and

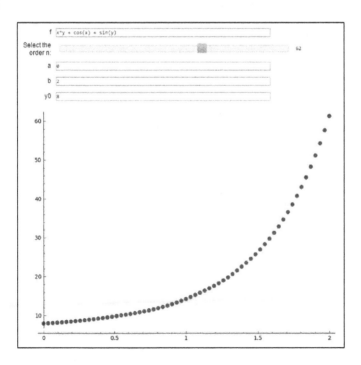

Another method that can be used for solving an IVP as the one given in formula (4.1) is using the **Picard Iteration**. Here is the method:

(1) We start with an initial solution: $y_0(x) = y_0$, for every x in $[a, b]$.

(2) Then, for $n \geq 1$, we recurrently compute

$$y_{n+1}(x) = y_0 + \int_a^x f(t, y_n(t))dt.$$

Under favorable conditions y_n will converge to the exact solution y of the IVP given in (4.1).

Here is our Sage Interact that follows this method:

```
var('x,t')
y = function('y',x)

@interact
def PicardDEInteract(
    f = input_box(default= "(x+y)^2"),
    n = slider(vmin=0, vmax=10, step_size=1,
               default=3,
               label="Select the order n: "),
    #the interval [a,b]
    a = input_box(default= 0),
    b = input_box(default = 5),

    #the initial condition for y(a)=y0:
    y0 = input_box(default = 2)):

    #this it to avoid a warning message
    f(x,y)=f

    #the approximate solution:
    sol(x) = y0
    print "n =", 0, "y(x)=", sol(x)

    #compute the step size
    for step in [1,2,..,n]:
        sol(x)=y0+integral(f(t, sol(t)), t, a, x)
        print "n =", step, "y(x)=", sol(x)
```

We obtain the following (aside from an error that disappears once you update the value of *n* using the slider provided):

```
  f  (x+y)^2

Select the
 order n:                                                                    2

    a  0

    b  5

   y0  2

  n = 0 y(x)= 2
  n = 1 y(x)= 1/3*x^3 + 2*x^2 + 4*x + 2
  n = 2 y(x)= 1/63*x^7 + 2/9*x^6 + 22/15*x^5 + 16/3*x^4 + 11*x^3 + 10*x^2 + 4*x + 2
```

and

```
  f  (x+y)^2

Select the
 order n:                                                                    6

    a  0

    b  5

   y0  2

  n = 0 y(x)= 2
  n = 1 y(x)= 1/3*x^3 + 2*x^2 + 4*x + 2
  n = 2 y(x)= 1/63*x^7 + 2/9*x^6 + 22/15*x^5 + 16/3*x^4 + 11*x^3 + 10*x^2 + 4*x + 2
  n = 3 y(x)= 1/59535*x^15 + 2/3969*x^14 + 272/36855*x^13 + 194/2835*x^12 + 23014/51975*x^11 + 3284/1575*x^10 + 254/35*x^9 +
  n = 4 y(x)= 1/109876902975*x^31 + 2/3544416225*x^30 + 7708/445414972275*x^29 + 109/313451775*x^28 + 72327172/1411943520487
  n = 5 y(x)= 1/760594829864786522589375*x^63 + 2/120729338073775638506250*x^62 + 92996/895616963768586567331687*x^61 + 1265
  n = 6 y(x)= 1/734700708925645316486517377626686846838667587109375*x^127 + 2/578504495217043556288598360335973597154862890052
```

Tweaking the previous code, one can obtain plots of the approximating solutions obtained:

```
var('x,t')
y = function('y',x)

@interact
def PicardDEInteract(
    f = input_box(default= "(x+y)^2"),
    n = slider(vmin=0, vmax=10, step_size=1,
               default=3,
               label="Select the order n: "),
    #the interval [a,b]
    a = input_box(default= 0),
    b = input_box(default = 2),

    #the initial condition for y(a)=y0:
    y0 = input_box(default = 1)):

    #this it to avoid a warning message
    f(x,y) =f
```

```
#the approximate solution:
sol(x) = y0
p = plot( sol(x), x, a,b, color = 'purple')

#compute the step size
for step in [1,2,..,n]:
    sol(x)=y0+integral(f(t, sol(t)), t, a, x)
    p += plot( sol(x), x, a,b,
               rgbcolor=hue(1/step),
               legend_label = str(step))

p.show()
```

We get

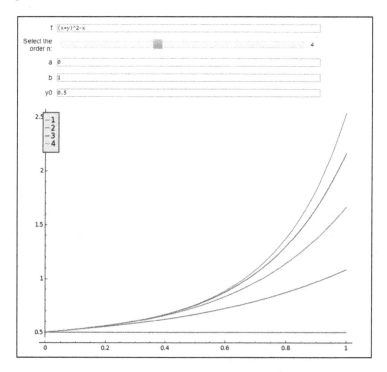

which plots the numerical approximations for the solution of

$$\begin{cases} y'(x) = (x+y)^2 - x \\ \quad\quad y'(0) = 0.5 \end{cases}$$

on $[0, 1]$.

4.12.1 Exercises

(1) Modify the given Sage Interact so that it allows the user to enter the exact solution (if known) and plot it. This would allow the user to visually inspect how good the table of approximations is compared to the exact solution.

(2) Use the Sage library function `eulers_method` to simplify the Sage Interact given in this section.

(3) Using a Numerical Analysis book as a guide, implement Sage Interacts for solving a system of differential equations, with initial conditions using Euler's Method.

(4) Using a Numerical Analysis book as a guide, implement Sage Interacts for solving a system of differential equations, with initial conditions using Taylor Series Method.

(5) Modify the Sage Interact given to numerically solve Differential Equations (using Euler Method) so that it displays the computed table instead of plotting it.

(6) Modify the Sage Interact given to numerically solve Differential Equations (using Euler Method) so that it displays the computed table and plots it.

(7) Using a Numerical Analysis book as a guide, implement Sage Interacts to numerically solve differential equations using Taylor Series Method. Your code should plot the table of solutions.

(8) Using a Numerical Analysis book as a guide, implement Sage Interacts to numerically solve differential equations using Taylor Series Method. Your code should plot the table of solutions and display the table of values computed.

(9) Using a Numerical Analysis book as a guide, implement Sage Interacts that plots in the same graph both

- the table of solutions of a differential equation that is obtained using Euler Method
- the table of solutions of a differential equation that is obtained using Taylor Series Method.

(10) Using a Numerical Analysis book as a guide, implement Sage Interacts that outputs a side-by-side table with the following values:

- the ones obtained using Euler Method
- the ones obtained using Taylor Series Method.

4.13 NUMERICAL METHODS FOR PARTIAL DIFFERENTIAL EQUATIONS

As mentioned earlier in this book, Sage's (but also SciPy and NumPy libraries') capabilities for solving ODE and PDE are currently quite limited. Therefore, one could choose to implement their own functions for doing numerical computations. Here we create a Sage Interact for solving the **Heat Equation Model problem**

$$\begin{cases} \frac{\partial^2}{\partial x^2} u(x,t) = \frac{\partial}{\partial t} u(x,t) \\ u(0,t) = u(1,t) = 0 \quad \text{for } x \in [0,1], t \geq 0 \\ u(x,0) = f(x) \end{cases}$$

using the **Finite-Difference Method**

$$\frac{1}{h^2}[u(x+h,t) - 2u(x,t) + u(x-h,t)] = \frac{1}{k}[u(x,t+k) - u(x,t)].$$

Using these equations, one can advance step by step, and obtain (as in the previous chapter) a table of values that will approximate the exact solution of the PDE under consideration (the Heat Equation Model in our case).

For more mathematical details about this topic, the reader is invited to read a Numerical Analysis book, such as [7].

Here comes the strategy for this method:

(1) **Step 0**: Choose values of k and h such that $k \leq \frac{h^2}{2}$ (e.g., $h = 0.1$, $k = 0.005$). Then, $n = \frac{1}{h}$.

(2) **Step 1**: Using the definition of $f(x)$, one can compute and print the boundary values:

$$u(0,0) = 0,$$
$$u(h,0) = f(h),$$
$$u(2h,0) = f(2h),$$
$$\vdots$$
$$u((n-1)h,0) = f((n-1)h),$$
$$u(1,0) = 0.$$

(3) **Step 2**: For each $t = k, 2k, \cdots, mk$:

- recursively compute $u(ih,t)$, for $i = 1, \cdots, n-1$, using

$$u(0,t) = u(1,t) = 0,$$
$$u(ih,t) = \frac{k}{h^2} u((i+1)h, t-k)$$
$$+ \left(1 - \frac{2k}{h^2}\right) u(ih, t-k)$$
$$+ \frac{k}{h^2} u((i-1)h, t-k).$$

Here comes the Sage Interact for this algorithm:

```
@interact
def PDEHeatInteract(
  h = input_box(default = 0.1),
  k = input_box(default = 0.005)):

    if(k>h^2/2):
        print "values not recommended to be used!"
    else:
        @interact
        def IterationsInteract(
            m = input_box(default = int(h/k)),
            f = input_box(default = sin(pi*x))):
            n = int(1/h)

            #avoid a warning:
            f(x)=f

            #initialize the array of values with 0's
            #this will be of size  (n+1) x (m+1)
            values = [[0 for col in [0,1,..,m]] for row in [0,1,..,n]]

            #step 1 ... initialize the first row
            for  row in [0,1,..,n]:
                values[row][0] = f(col*h)

            #step 2
            for t in [1, 2, .., m]:
                values[0][t] = 0
                values[1][t] = 0

                for i in [1, 2,.., n-1]:
                    values[i][t] =( k/h^2*values[i+1][t-1]\
                                 + (1-2*k/h^2)*values[i][t-1] \
                                 +k/h^2*values[i-1][t-1]).n()

            #IMPORTANT!!! for integer indices, scaling occurred
            #values[i,j] should be read as sol(i*h, j*k)!!!

            #we use the option colorbar = true
            #     in order to get the vertical bar on the right
            matrix_plot(values,colorbar=True, cmap='Oranges',\
                        aspect_ratio='automatic').show()
```

We obtain

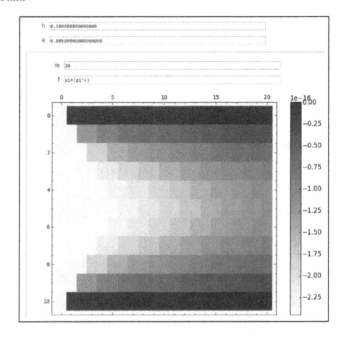

and

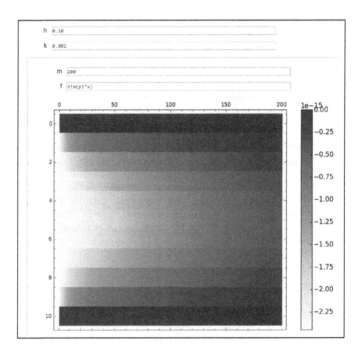

Choosing the following values $h = k = 0.01$, we obtain an error message:

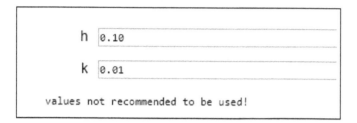

Changing some of the values (similar to the following screenshots), we obtain

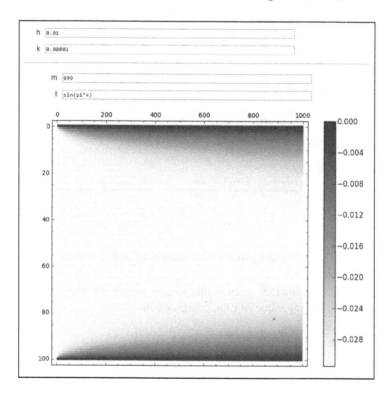

and

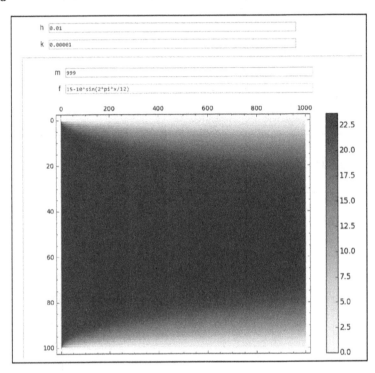

4.13.1   Exercises

(1) Change the Sage Interact given in this section so that it prints out the obtained table of values.

(2) Change the Sage Interact given in this section so that it allows the user to pick one of the following values for the cmap option:
   - 'Accent'
   - 'cool'
   - 'gnuplot'
   - 'ocean'
   - 'prism'
   - 'Purples'
   - 'Spectral'
   - 'terrain'
   - 'winter'.

(3) Using a Numerical Analysis book, create a Sage Interact similar to the one given in this section that implements the Crank–Nicholson Method for numerically solving the heat equation.

(4) Using a Numerical Analysis book, create a Sage Interact similar to the one given in this section that implements the Lax–Friedrichs Method for the solution of hyperbolic PDEs.

4.14 SCATTER PLOTS – LINE OF BEST FIT AND MORE

Given a set of data points, one can obtain a scatter plot one can use Sage code similar to the one below:

```
points = [ (0, 1.1), (1, 2.1), (2, 5.1), (3, 9.9), (4, 14) ]
scatter_plot( points )
```

to obtain

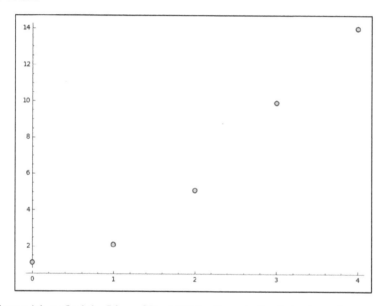

If we wish to find the **Line of Best Fit** for the set of points given above, one can use the following Sage code:

```
points = [ (0, 1.1), (1, 2.1), (2, 5.1), (3, 9.9), (4, 14) ]

var('a,b')
model(x) = a*x+b
find_fit(points, model)
```

We obtain

```
[a == 3.3600000000051473, b == -0.2800000000027889]
```

Next, we construct a Sage Interact that allows user to input a set of data points and then finds and plots the Line of Best Fit along with a scatter plot of the given set of points.

```
var('a,b')

@interact
def LineFitInteract(
    points = input_box(default = [ (0, 1), (1, 2),\
                        (2, 5), (3, 9), (4, 14) ])):
    var('x')
    model(x) = a*x+b
    #compute the range
    xmin = min([points[i][0] for i in [0,1,..,len(points)-1]])
    xmax = max([points[i][0] for i in [0,1,..,len(points)-1]])
    p   = scatter_plot( points )

    fit = find_fit(points, model)
    f(x) = model(a =fit[0].rhs(), b = fit[1].rhs())
    p += plot(f(x), xmin, xmax )

    show(p)
```

We obtain

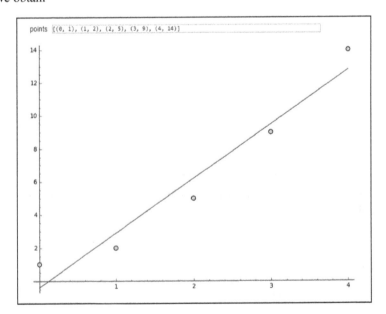

We can give the list of points as a `for` loop as it shows in the next image:

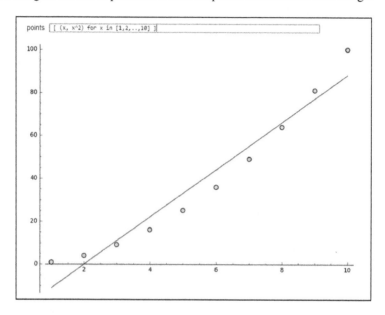

Adding `show(f(x))` right after the last line of the Sage Interact, we obtain the expression of the best fit function:

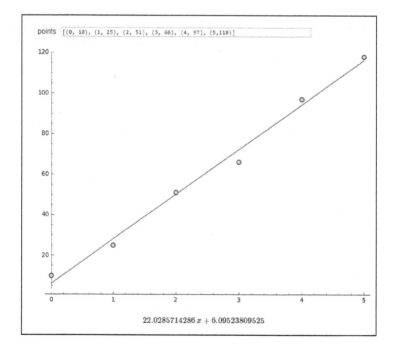

Using the following input table `[(x+random()/10, x+random()/10)` `for x in [0, 0.01,..,1]]`, we obtain

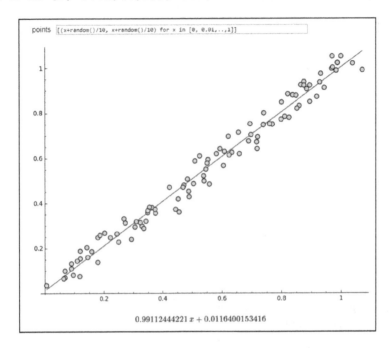

The following Sage Interact allows the user to input any Mathematical models, such as a line ($ax + b$), a quadratic ($ax^2 + bx$), or any other models that involved exactly two parameters a, b :

```
var('a,b,x')
@interact
def LineFitInteract(
    points = input_box(default = [ (0, 1), (1, 2),\
                       (2, 5), (3, 9), (4, 14) ]),
    model = input_box( default = a*x+b )):
    var('x')
    model(x) = model
    #compute the range
    xmin = min([points[i][0] for i in [0,1,..,len(points)-1]])
    xmax = max([points[i][0] for i in [0,1,..,len(points)-1]])
    p   = scatter_plot( points )

    fit =  find_fit(points, model)
    f(x) = model(a =fit[0].rhs(), b = fit[1].rhs())
    p += plot(f(x), xmin, xmax )
    show(p)
    show(f(x))
```

We obtain, for example, for $ax^2 + bx$:

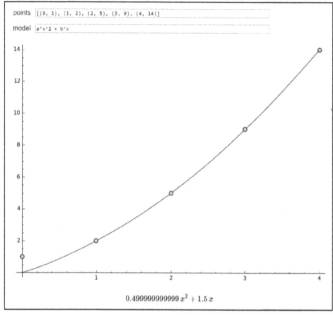

and for $ax^2 + b$

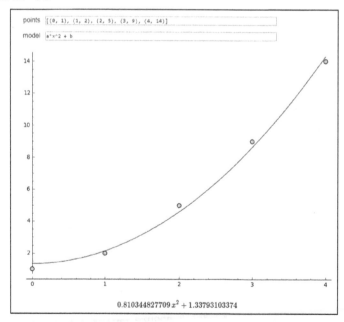

The following Sage Interact allows the user to input two Mathematical models, such as a line $(ax + b)$ and a quadratic $(ax^2 + bx + c)$ or any other models that involve

two to four constants types of terms a, b, c, d. One can easily extend this code to accept
more complicated models.

```
var('a,b,c,d,x')
@interact
def LineFitInteract(
    points = input_box(default = [ (0, 1), (1, 2),\
                        (2, 5), (3, 9), (4, 14) ]),
    model1 = input_box( default = a*x^2+b*x+c ),
    model2 = input_box( default = a*cos(x)+b*sin(x) )):
    var('x')
    model1(x) = model1
    model2(x) = model2
    #compute the range
    xmin = min([points[i][0] for i in [0,1,..,len(points)-1]])
    xmax = max([points[i][0] for i in [0,1,..,len(points)-1]])

    #produces the scatter plot
    p  = scatter_plot( points )

    fit1 =  find_fit(points, model1)
    if( len(model1) == 2 ):
        f1(x) = model1(a =fit1[0].rhs(), b = fit1[1].rhs())

    elif( len(model1) == 3 ):
        f1(x) = model1(a =fit1[0].rhs(), b = fit1[1].rhs(), \
                        c = fit1[2].rhs())

    elif( len(model1) == 4 ):
        f1(x) = model1(a =fit1[0].rhs(), b = fit1[1].rhs(), \
                        c = fit1[2].rhs(), d = fit1[3].rhs())
    else:
        raise RuntimeError("model 1 should have 1-4 parameters!")
    p += plot(f1(x), xmin, xmax, \
                color = 'red',       \
                legend_label = "model1 =$"+str (f1(x))+"$" )


    fit2 =  find_fit(points, model2)
    if( len(model2) == 2 ):
        f2(x) = model2(a =fit2[0].rhs(), b = fit2[1].rhs())

    elif( len(model2) == 3 ):
        f2(x) = model2(a =fit2[0].rhs(), b = fit2[1].rhs(), \
                        c = fit2[2].rhs())

    elif( len(model2) == 4 ):
        f2(x) = model2(a =fit2[0].rhs(), b = fit2[1].rhs(), \
                        c = fit2[2].rhs(), d = fit2[3].rhs())
    else:
        raise RuntimeError("model 2 should have 1-4 parameters!")
    p += plot(f2(x), xmin, xmax, \
                color = 'blue',      \
                legend_label = "model2 =$"+str (f2(x))+"$")
    show(p)
```

We obtain

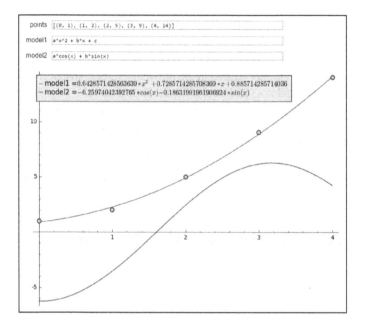

and for the set of points `[(x,sin(x)) for x in [0,pi/4,..,2*pi]]`

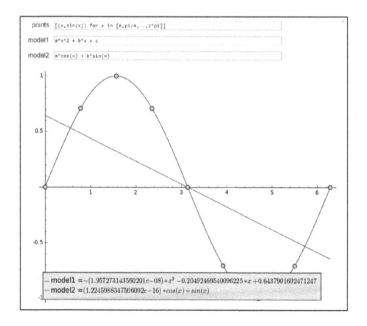

4.14.1   Exercises

(1) Create a Sage Interact that allows the user to input a set of data points and to select
(using a selector) a model from the following set: $y = ax + b$, $y = ax^2 + bx + c$,
$y = a\cos(x) + b$, and $y = a * \cos(x) + b * \sin(x)$. Then, the interact should display the model, and it should plot the model and the given set of data points.

(2) Expand the Sage Interact created in the previous exercise so that it will allow the
user to select a thickness and a color for the fitting curve.

(3) Expand each of the Sage Interacts given in this section so that they allow the user
to choose to either plot the graph or create a png file that can be downloaded.
Hint: Refer to the example given on p. 29.

(4) Expand the Sage Interacts given above so that the user can select to either have
the graphs displayed or have them saved into either a 'png' or a 'pdf' file. Hint:
Refer to the example given on p. 29.

4.15   MATRICES, EIGENVALUES, AND EIGENVECTORS

Given a matrix A, one can compute its eigenvalues using Sage code similar to the one
below:

```
A = matrix([[1,1,0],[1,1,0],[2,4, 3]])
eigenVals = A.eigenvalues()
print eigenVals
```

and obtain `[3, 2, 0]`.

The following Sage Interact will allow the user to select a value of n and it will
randomly generate an $n \times n$ matrix with integer elements. Then it will compute and
display the **eigenvalues** of the matrix.

```
@interact
def MatrixInteract(
    n = slider(vmin=1,vmax=50,
               default = 2,
               step_size = 1)):
    #create a random nxn matrix
    A = random_matrix(ZZ, n)

    #display the random matrix A
    show(A)

    #compute and print its eigenvalues
    print  A.eigenvalues()
```

We obtained

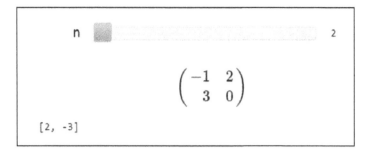

n   2

$$\begin{pmatrix} -1 & 2 \\ 3 & 0 \end{pmatrix}$$

`[2, -3]`

and by selecting a larger value of *n*, we get

n   3

$$\begin{pmatrix} 4 & 5 & 1 \\ 1 & 0 & -1 \\ 0 & 0 & 4 \end{pmatrix}$$

`[5, 4, -1]`

Note: The eigenvalues are not always integers!

To let Sage compute the eigenvalues, and their corresponding **eigenvectors**, along with the number of their multiplicities, one simply has to replace the last line in the previous Sage Interact with

```
print  A.eigenvectors_left()
```

and obtain

n   2

$$\begin{pmatrix} 2 & 0 \\ 1 & -1 \end{pmatrix}$$

```
[(2, [
(1, 0)
], 1), (-1, [
(1, -3)
], 1)]
```

meaning the obtained matrix has the eigenvalues 2 (multiplicity 1, eigenvector $(1,0)$), and -1 (multiplicity 1, eigenvector $(1, -3)$).

To reduce a matrix into its corresponding **echelon form**, one can replace the last line with

```
print   A.echelon_form()
```

or

```
print   A.rref()
```

and obtain

$$n \boxed{} \qquad 2$$

$$\begin{pmatrix} 1 & -3 \\ -2 & -5 \end{pmatrix}$$

```
[ 1  8]
[ 0 11]
```

and

$$n \boxed{} \qquad 6$$

$$\begin{pmatrix} 1 & -1 & 0 & 0 & 0 & 20 \\ -1 & -1 & 0 & 3 & 0 & 1 \\ 5 & 2 & 0 & 3 & 0 & -10 \\ 0 & 0 & 7 & 2 & 3 & 0 \\ 5 & 1 & 1 & -1 & -1 & 4 \\ 2 & 0 & -1 & 0 & -3 & -5 \end{pmatrix}$$

```
[ 1  0  0  0  0 511]
[ 0  1  0  0  0 491]
[ 0  0  1  0  1 828]
[ 0  0  0  1  0  31]
[ 0  0  0  0  2 199]
[ 0  0  0  0  0 910]
```

One can also use the SciPy package in order to find the eigenvalues of a given matrix. For example,

```
import scipy
from scipy.linalg import eigvals
@interact
def MatrixInteract2(
    n = slider(vmin=1,vmax=50,
               default = 2,
               step_size = 1)):
    #create a random nxn matrix
    A = random_matrix(ZZ, n)
        #display the random matrix A
    show(A)
        #compute and print its eigenvalues
    print  scipy.linalg.eigvals(A)
```

will output something similar to

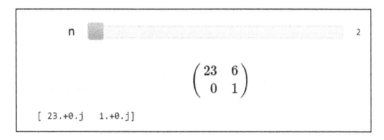

$$\begin{pmatrix} 23 & 6 \\ 0 & 1 \end{pmatrix}$$

```
[ 23.+0.j   1.+0.j]
```

and

```
n                                              10
```

$$\begin{pmatrix}
0 & 2 & -2 & 1 & -25 & 5 & 1 & 0 & 0 & -1 \\
-2 & 0 & -2 & 2 & -3 & 3 & -1 & 1 & -4 & 9 \\
-1 & 3 & -12 & 1 & -4 & -1 & -1 & -1 & 0 & -19 \\
-2 & 1 & -1 & 3 & 1 & 0 & 0 & 0 & 6 & -1 \\
4 & 5 & -28 & -1 & 0 & 4 & -1 & 2 & 81 & -3 \\
-6 & 1 & -1 & 0 & 1 & 0 & 0 & 1 & 4 & -1 \\
1 & 0 & 8 & -3 & 2 & 1 & 0 & 1 & 1 & -1 \\
-1 & 2 & 2 & 1 & -5 & -1 & 0 & -1 & -1 & -1 \\
0 & 90 & -1 & 16 & 0 & -1 & -1 & -2 & 1 & 1 \\
0 & -1 & 0 & 2 & 0 & -1 & -12 & 10 & 0 & 4
\end{pmatrix}$$

```
[   6.87487634+29.32233953j   6.87487634-29.32233953j -29.13041604 +0.j
   13.65862102 +1.47480607j  13.65862102 -1.47480607j
   -5.10227528+11.30223284j  -5.10227528-11.30223284j -10.09868732 +0.j
   -1.00816428 +0.j           4.37482348 +0.j          ]
```

The following Sage Interact can be used to compute the **Cholesky decomposition of a matrix**, the **QR decomposition of a matrix**, the **pivoted LU decomposition of a matrix**, and the **Hessenberg form of a matrix**. In order to keep the matrix fixed between different choices of decompositions, we used nested interacts.

```
import scipy
from scipy.linalg import cholesky
from scipy.linalg import qr
from scipy.linalg import lu
from scipy.linalg import hessenberg

@interact
def DecompositionInter(
    n = slider(vmin=1,vmax=50,
               default = 2,
               step_size = 1)):
    #create a random nxn matrix
    A = random_matrix(ZZ, n)

    #display the random matrix A
    show(A)

    @interact
    def InnerInteract(
    decomposition = selector(
            values = ["cholesky", "qr", "lu", "hessenberg"],
            default = "hessenberg")):
        if(decomposition == "cholesky"):
            print  scipy.linalg.cholesky(A)
        elif(decomposition == "lu"):
            print  scipy.linalg.lu(A)
        elif(decomposition == "qr"):
            print  scipy.linalg.qr(A)
        else:
            print  scipy.linalg.hessenberg(A)
```

We obtained

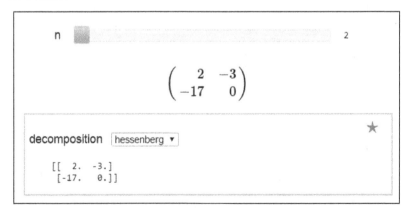

and

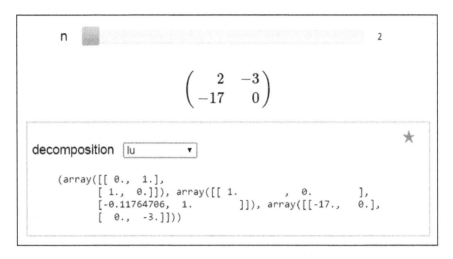

Selecting another matrix and another decomposition we obtained

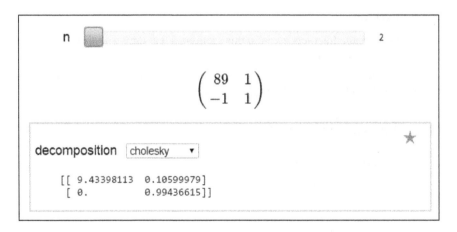

If one changes the very last line of code of the previous Sage Interact to

```
show( matrix( scipy.linalg.hessenberg(A)) )
```

we obtain a much nicer output:

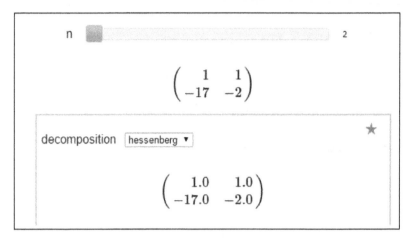

We leave it as an exercise to change all the other outputs (for QR, for LU, and for Cholesky) to such a nice output.

4.15.1   Exercises

(1) Change all Sage Interacts given in this section so that they will allow the user to input the matrix into an input box.

(2) Change all Sage Interacts given in this section so that they will use the show method instead of print to display the results.

4.16   SOLVING MATRIX EQUATIONS

To solve a matrix equation such as

$$AX = Y,$$

where A is a matrix, X, Y are vectors, one can use the method solve_right() as shown in the following Sage Interact:

```
import random

@interact
def MatrixEqInteract(
    n = slider(vmin=1,vmax=50,
               default = 2,
               step_size = 1)):
```

```
#create a random nxn matrix
A = random_matrix(ZZ, n)
Y = vector([random.randint(1,10) for i in [1,2,..,n]])

#display the random matrix A
print "Solving AX=Y, where"
print "A = "
show(A)
print
print "Y = ",
show(Y)
print
print
print "Solution = "
show(  A.solve_right(Y)  )
```

We obtained

```
n                                          2

Solving AX=Y, where
A =
```

$$\begin{pmatrix} -2 & -1178 \\ -2 & 0 \end{pmatrix}$$

```
Y =
```

$$(9, 1)$$

```
Solution =
```

$$\left(-\frac{1}{2}, -\frac{4}{589}\right)$$

and for $n = 7$,

```
n        [   ]                                      7

Solving AX=Y, where
A =
```

$$\begin{pmatrix} 0 & 0 & -1 & -1 & 12 & 2 & 0 \\ 1 & -1 & -1 & 1 & -2 & 9 & -2 \\ 1 & -1 & 1 & 1 & -1 & -1 & 11 \\ 0 & -13 & 1 & 1 & 1 & -1 & 1 \\ 0 & -1 & 1 & 0 & 3 & -1 & 5 \\ 1 & -5 & 1 & -1 & -1 & -1 & -1 \\ -17 & 1 & 1 & 2 & 1 & -1 & 0 \end{pmatrix}$$

```
Y =
```

$$(3, 1, 2, 3, 8, 6, 2)$$

```
Solution =
```

$$\left(\frac{22679}{44999}, \frac{43043}{89998}, \frac{440431}{44999}, \frac{14631}{44999}, \frac{39908}{44999}, \frac{6539}{5294}, -\frac{24612}{44999} \right)$$

Note: The matrix A does not have to be square! For example,

```
A = matrix([[1, 0, 0], [0, 2, 0], [0, 0, 1], [1, 2, 1]])
Y = vector([10, 20, 30, 60])
print "Solution = "
show(   A.solve_right(Y)   )
```

outputs

```
Solution =
```

$$(10, 10, 30)$$

4.16.1   Exercises

(1) Change the Sage Interact given in this section so that it will allow the user to input the matrix and the vector using input boxes.

(2) Create a Sage Interact that allows the user to input matrices A, B, and C, and then it displays the solution matrix of the equation AX+B=C (if it exists!).

REFERENCES

1. G.A. Anastassiou, I. Iatan, *Intelligent Routines II: Solving Linear Algebra and Differential Geometry with Sage*, Springer, 2014.

2. G.A. Anastassiou, R.A. Mezei, *Numerical Analysis Using Sage*, Springer, 2015.

3. R. Bannon et al., *Embedding Sage in LATEX with SAGETEX for TEXShop Users*, 2014, http://faculty.essex.edu/~bannon/Sage_cs/sagetextexshop.pdf.

4. G.V. Bard, *Sage for Undergraduates (online version)*, 2015, http://www.gregorybard.com/sage.html.

5. R.A. Barnett, M.R. Ziegle, K.E. Byleen, *Finite Mathematics For Business, Economics, Life Sciences And Social Sciences 11th edition*, Pearson, 2010.

6. F.J. Blanco-Silva, *Learning SciPy for Numerical and Scientific Computing*, Packt Publishing, Birmingham, 2013.

7. W. Cheney, D. Kincaid, *Numerical Mathematics and Computing*, 7th edition, Brooks/Cole: Cengage Learning, Boston, MA, 2013.

8. J.F. Epperson, *An Introduction to Numerical Methods and Analysis*, revised edition, John Wiley & Sons, Inc., 2007.

9. T. Gaddis, *Starting out with Python*, 3rd edition, Addison-Wesley, 2014.

10. NumPy, 2015, http://docs.scipy.org/doc/numpy/reference.

11. M. O'Sullivan, D. Monarres, *SDSU Sage Tutorial*, 2011, http://www-rohan.sdsu.edu/~mosulliv/Teaching/sdsu-sage-tutorial/index.html.

12. T.J. Rivlin, *An Introduction to the Approximation of Functions*, Dover Publications, New York, 2003.

13. Sage Cell Server, http://sagecell.sagemath.org/, retrieved Sept. 2015.

An Introduction to SAGE Programming: With Applications to SAGE Interacts for Numerical Methods, First Edition. Razvan A. Mezei
© 2016 John Wiley & Sons, Inc. Published 2016 by John Wiley & Sons, Inc.

14. SciPy, 2015, http://docs.scipy.org/doc/scipy/reference.

15. W.A. Stein, *Sage for Power Users*, 2012, http://www.wstein.org/books/sagebook/sagebook.pdf.

16. W.A. Stein et al., *Sage Mathematics Software (Version 6.3), The Sage Development Team*, 2014, http://www.sagemath.org.

17. M. Sullivan, K. Miranda, *Calculus – Early Transcendentals*, W.H. Freeman and Company, 2014.

18. C. Warren, K. Denley, E. Atchley, *Beginning Statistics*, Hawkes Learning Systems, 2014.

INDEX

An Introduction to SAGE Programming: With Applications to SAGE Interacts for Numerical Methods,
First Edition. Razvan A. Mezei
© 2016 John Wiley & Sons, Inc. Published 2016 by John Wiley & Sons, Inc.